the wolf
IS AT THE DOOR

**HOW TO SURVIVE AND THRIVE
IN AN AI-DRIVEN WORLD**

BEN ANGEL

Entrepreneur Press®

Entrepreneur Press, Publisher
Cover Design: Andrew Welyczko
Production and Composition: Mike Fontecchio, Faith & Family Publications

© 2024 by Ben Angel
All rights reserved.
Reproduction or translation of any part of this work beyond that permitted by Section 107 or 108 of the 1976 United States Copyright Act without permission of the copyright owner is unlawful. Requests for permission or further information should be addressed to Entrepreneur Media, LLC Attn: Legal Department, 2 Executive Cir. #150, Irvine, CA 92614.

This publication is designed to provide accurate and authoritative information in regard to the subject matter covered. It is sold with the understanding that the publisher is not engaged in rendering legal, accounting, or other professional services. If legal advice or other expert assistance is required, the services of a competent professional person should be sought.

Entrepreneur Press® is a registered trademark of Entrepreneur Media, LLC

Library of Congress Cataloging-in-Publication Data

Names: Angel, Ben, author. | Entrepreneur Press, issuing body.
Title: The wolf is at the door : how to survive and thrive in an AI-driven world / by Ben Angel.
Description: Irvine, CA : Entrepreneur Press, 2024. | Includes index. |
  Summary: "A compelling and cautionary narrative about surviving the AI revolution and how to tune out the noise in a world full of distractions"—Provided by publisher.
Identifiers: LCCN 2023044855 (print) | LCCN 2023044856 (ebook) | ISBN 9781642011654 (hardback) | ISBN 9781613084762 (epub)
Subjects: LCSH: Artificial intelligence--Social aspects. | Social media—Psychological aspects. | Social media and society.
Classification: LCC HM851 .A6654 2024 (print) | LCC HM851 (ebook) | DDC 303.48/33—dc23/eng/20231117
LC record available at https://lccn.loc.gov/2023044855
LC ebook record available at https://lccn.loc.gov/2023044856

Printed in the United States of America

27 26 25 24   10 9 8 7 6 5 4 3 2 1

LEGAL DISCLAIMER. This book and its contents are intended for educational purposes only and should in no way be interpreted as medical or any other advice concerning the use of products or services featured in this book. This book reflects the author's personal experience with the products, services, and methods detailed herein, and does not constitute an endorsement or recommendation by publisher or its representatives of any of those products, services, or methods. In addition, the views, opinions, and advice contained in this book represent the views, opinions, and advice of the author and/or any third parties identified herein and not of the publisher or its representatives. The author is not a certified or licensed health-care professional or an authorized representative of any of the products, services, or methods mentioned in this book. Because the advice, strategies, and recommendations contained in this book may not be suitable for your situation, the author and publisher make no expressed or implied warranties, and assume no liability whatsoever, in connection therewith, including warranties about the type or extent of any benefits to be gained from attempting a similar regimen. The author and publisher recommend that anyone reading this book consult with appropriate licensed professionals and other experts before taking any action in connection with, or based on, the contents of this book.

The illustrations in this book were created using Playground, an AI-powered platform designed for generating digital images and graphics. These illustrations are subject to the license terms and conditions found at https://playgroundai.com/terms.

Contents

Introduction ix

Rule #1: Expect the Unexpected 1

> "The ring of the phone pierced through the quiet halls of Ruth's home in Regina, Saskatchewan."
> ADVENTURERS GUIDE

Rule #2: Avoid Cruel Optimism 17

> "As I stand disoriented and lost, the distant sound of sirens and crowds attracted my attention."
> ADVENTURERS GUIDE

Rule #3: Fuel Your Focus 39

> "As children, we weren't fed false hope; rather, we faced harsh realities."
> ADVENTURERS GUIDE

Rule #4: Open a Window 53

> "This wolf scratches at the door at night, keeping you awake and preventing you from finding your flow state, stealing your attention and peace of mind."
> ADVENTURERS GUIDE

Rule #5: Accelerate Adaptability 67

> "The West finds itself grappling with the challenges that will hinder our ability to adapt."
> ADVENTURERS GUIDE

Rule #6: Embrace Reconstruction 83

"Abruptly, around 2 a.m., a noise from the far side of the other bed jolts me awake. Now 53, my father struggles to communicate."
ADVENTURERS GUIDE

Rule #7: Find Your Frequency 99

"Let the hunger games begin!" I announce to a new friend. As they wish me well, I step into the room."
ADVENTURERS GUIDE

Rule #8: Boost Your Brain Power 117

"Suddenly, I see a trash can go flying across the empty Manhattan Street. "Oh no, it's too late," I think.
ADVENTURERS GUIDE

Rule #9: Master the Art of Intuition 135

"Fuck it, what's the worst that could happen?"
ADVENTURERS GUIDE

Rule #10: Make Tough Decisions, Fast 153

"I don't want him to die, but I also don't want him to suffer."
ADVENTURERS GUIDE

Rule #11: Know Who You Are 173

"Hasn't it dawned on you yet?"

Endnotes 187

Acknowledgments 211

About the Author 213

Index 215

Introduction

It was not long before the wolf arrived at the old woman's house. He knocked at the door: tap, tap. "Who's there?" "[It is] your grandchild, Little Red Riding Hood," replied the wolf, counterfeiting her voice, "who has brought you a cake and a little pot of butter sent you by mother.

—Charles Perrault, "Little Red Riding Hood"

THE RING OF THE PHONE pierced through the quiet halls of Ruth's home in Regina, Saskatchewan. The 73-year-old Ruth, with short gray hair framing her gentle face, picked it up to hear her grandson Brandon's panicked voice. He was calling from a prison cell. Wallet-less and in dire need of cash to make bail, he pleaded with his grandma to come up with a hefty sum of money—and fast. Sharing the shocking news with her husband Greg, 75, in a panic, they drove to their bank and withdrew 3,000 Canadian dollars. She was determined to do whatever it took to free her grandson as quickly as possible. But $3,000 wasn't enough! Determined to succeed, Ruth and Greg drove to a second bank, where they shared the dire news with the bank manager, hoping to gain help in withdrawing additional funds. It was only then that the second bank manager's suspicions were piqued. The shocking revelation: Another unsuspecting customer had received a similar call and uncovered the chilling truth: The unnervingly realistic voice had been faked—the man on the phone wasn't their grandson, but a wolf in disguise. "It was definitely this feeling of . . . fear," Ruth revealed to *The Washington Post*.[1]

The wolf is no longer knocking at your door. He has kicked it open, stolen your attention, and now he's coming for your wallet. If you think this is hyperbole, you haven't been paying attention. Ruth's story isn't the first, nor will it be the last of its kind. It evokes vivid memories of the wolf from "the beloved fable Little Red Riding Hood"—a cautionary tale whispered by parents for centuries to warn their children of the world's harsh realities. Only this time, the wolf, cunning and sly, didn't waste time knocking on Grandma's door with the promise of cake and butter. Instead, he just picked up the phone and dialed, hitting at the heart of what it means to be human:

INTRODUCTION

connection. His malicious intent wasn't to devour his prey, but to bleed their bank account dry. Even though he didn't succeed in that goal, he accomplished something much more sinister: shattering our sense of reality as we know it.

When we hear a loved one's voice on the phone, we expect it to be them, not an artificially generated (AI) version of them that can manipulate us into drawing down our life's savings or, worse, committing crimes. With AI's newfound capabilities, a few seconds of audio, an unsuspecting target, and the powerful bond between two people willing to risk everything to protect each other at any cost is enough to set the stage. Only this time, you may find yourself unwillingly written into the story line.

New developments in AI have surpassed our brain's capacity to process and conceptualize what a world with AI looks like, as well as our emotional capability to grasp its far-reaching impact. Historically, the wolf has represented chaos and destruction—hence the phrase "keep the wolf from the door" as an embodiment of the struggle against economic upheaval. The Great Depression of the 1930s saw people fighting to keep the wolf at bay as they lost jobs and homes. During the Great Recession of 2007-2009, the wolf lurked in the shadows of the financial collapse, with tent cities springing up across America as nearly nine million workers found themselves jobless.[2] Today, the wolf has emerged yet again, armed with new tools to exploit our most significant vulnerabilities: The need for a purpose, job security, and our grip on reality as we know it. Investment bank Goldman Sachs has estimated that AI could automate nearly 300 million jobs,[3] Alarmingly, we are wildly unprepared for both the dangers and the opportunities.

Outdated self-help, productivity, and time management strategies pale in comparison to the addictive nature of these new technologies. AI's brilliant yet alarming potential to empower and destabilize us elicits visceral reactions that we are struggling to process. We are clumsily making our way into uncharted territory, and it's time for a new approach to success to tap into the greatest competitive advantage of the 21st century: An unrivaled ability to focus in a world full of distractions. But with it, we must combat a threat that will usher in changes to our society faster than the Industrial Revolution.

The leaders of the future must be able to tune out the noise so they can tune into what matters most while keeping stress and burnout at bay. They will hold the lion's share of business and personal opportunities. They will possess zero limits, but they will need to compete at a level we've never seen before against technology evolving to be smarter, faster, and cheaper than they could ever be. They will need to upgrade their brains to adapt to a rapidly changing landscape that has captured our imaginations and forced us to reconceptualize reality itself. The question is, how?

Over the years, I've anticipated future trends, embracing ideas that once seemed outlandish. In 2009, I championed personal branding by speaking at more than 60 business events years before the topic gained significant traction. In 2013, I foresaw the online education boom in my book *Flee 9 to 5*. Back then, during a live TV interview, I faced ridicule as the host compared my ideas to scam ads on power poles. Now, the industry is projected to reach $602 billion by 2030.[4] In 2018, I embarked on a 90-day mission to biohack my way back to health after battling depression. I explored the use of wearable devices, smart drugs, and nutritional supplements tested by the military in my bestselling book *Unstoppable*. I'm not a doctor, so I was humbled when *Unstoppable* sold more than 80,000 copies and won praise from doctors, psychiatrists, and nutritionists for its groundbreaking approach to mental health. In 2021, I delved into nutritional psychology and gut microbiome research in my book *Mind Control*. I spoke with nutritional psychologists and gut health experts, to investigate the gut-brain axis and how what we eat, along with the millions of gut bacteria we carry, can influence our behavior.

As we venture further into the 21st century, I've turned my attention to AI, the next frontier in innovation. With its unprecedented potential to revolutionize entire industries and disrupt established norms, AI

INTRODUCTION

mirrors humanity's best—and worst. This has led to increased demand for innovative approaches to help individuals and organizations remain competitive while holding onto a semblance of work/life balance that has been hanging by a thread for decades. It's also ignited a need to anticipate emerging trends for success while navigating countless threats. Opinions diverge on AI's impact, from dystopian futures with job automation to life-extending medical advances. The question remains: How will it reshape our lives, careers, and businesses in the coming years? And, more important, how do we wrap our heads around these developments when we must constantly question whether what we see, hear, or read is real?

In this book, we will dial in on when this disruptive change began while taking a long-term view of what the future may hold. We will uncover advanced strategies to upgrade our brains and way of thinking, not just to cope but to thrive during this extraordinary period. My aspiration is unambiguous: AI got an upgrade, which means we must, too—not just to collaborate or compete with it, but also to ensure we have the mental bandwidth to adapt to it.

We will examine the threats and opportunities that these new developments present in every aspect of our lives: employment, finance, grief, behavior, sex, normalcy, and even death, with a particular focus on the social, ethical, and financial dilemmas that arise. These advances offer us a chance to streamline our lives, prepare for the future, identify vulnerabilities, and supercharge our productivity and profits.

We'll delve into the fragility of the human experience and explore how technology and nutrition offer opportunities to enhance ourselves without resorting to extremes like Elon Musk's brain implants.[5] Throughout this journey, we will discover our strength, encourage critical and imaginative thinking, and offer a framework for how to approach and use these technologies at the end of each chapter in your "Adventurer's Handbook." This will help you reimagine your future and act as an essential guide for navigating this ever-changing landscape.

But let me be clear: I am not an expert in AI. I am an author with almost two decades of experience in studying a variety of topics by meticulously examining the available research, speaking to the experts in relevant fields, testing various strategies, and then assembling it all into an easy-to-follow plan. My background equips me with the ability to present complex topics in an easy-to-understand manner. This means I will not drown you in tech talk, and I will not pretend to have all the answers: no one does. But what I will do is open your eyes to a new and unfamiliar world and refamiliarize you with the tools you already possess so you can use them to adapt in unexpected ways.

Change is inevitable, but it is also cruel and doesn't care who you are or what position you hold. *Unstoppable* covered little-known modern-day threats to our mental health and gave tens of thousands of readers a roadmap for optimizing every aspect of their lives. That journey continues as we face the next challenge by forging a new path through uncharted territory. This time, the wolf is no longer a character from a fairy tale but a possible threat looming in the shadows of your future. This story is unwritten, the path untrodden. When you meet the wolf, will he be friend or foe? By the end of this book, you'll be able to decide whether you'll tame the beast when you encounter him—or ignore his existence at your own peril.

RULE #1

Expect the Unexpected

REALITY STRIKES as the driver heaves my bulky luggage from the trunk of his iconic New York yellow taxi and drops it on the ground. I find myself in a dimly lit alley, frantically trying to connect to the internet to locate my hotel. As I stand disoriented and lost, the distant sound of sirens and crowds attracts my attention. Glancing up, I see the bright lights of Times Square illuminating the intersection down the street. At 90 seconds to midnight in October 2012, I draw on the last of my reserves after a long-haul flight from Australia and plunge into the madness. As I approach, the sense of overwhelm intensifies in my chest. The thousands of billboards adorning the towering skyscrapers take me by surprise. "You can do this," I whisper, reassuring myself. I inhale deeply and search for the nearest Starbucks (and its Wi-Fi), thinking, "What the fuck have I gotten myself into!?"

Driven by pure ambition, I had enrolled in a media summit to pitch my next book to dozens of top TV producers from shows like *Good Morning America* and *Today*, as well as book publishers in the Big Apple. It seemed like an incredible opportunity. My tenacity, inherited from my farmer dad, told me, "If it doesn't make you nervous, it's not worth doing." Australia was home, but I craved a challenge that would stretch my limits. If I could find that anywhere, it would be Manhattan.

Desperately, I try to focus through the noise and lights that are assaulting me from every direction. If I am to adapt to this new environment, I must be fast. I have less than 36 hours to rest, recover from my jet lag, prepare for the summit, and get to the historic Roosevelt Hotel. Scanning my surroundings, I finally spot a bustling Starbucks jam-packed with noisy tourists and a line out the door. I make a beeline toward it, pushing through the crowd, and try to connect to their Wi-Fi through the window. My relief is short-lived when I discover an email from my assistant back home: "Ben, your Airbnb was canceled while you were in transit; you're not staying in Times Square anymore. Your new hotel is . . ." I let out a sigh of exasperation and snap a screenshot of the new hotel's information on my painfully slow iPhone 4.

    I lug my heavy bags to the curb, where I quickly learn that hailing a cab in New York means assertively staking your claim and being clear on where you are going. Little did I know that this was the first of many anxiety-inducing tests on this trip that would rip the rug out from under my feet and challenge my ability to adapt. My most significant trial lay just a few days ahead—one that would dominate international headlines, catch millions off-guard, put me in harm's way, and tragically result in the deaths of more than a hundred people.

    Years before I hit the streets of Manhattan, from the ages of 23 to 30, I hosted monthly business-to-business networking events, conducting countless interviews with experts to keep my members at the forefront of innovation and technology. But as 2022 drew to a close, for me, as for many others, I found the daily deluge of AI news both exhilarating and terrifying. Scouring thousands of articles for guidance, experimenting with a multitude of AI tools, and viewing the heated exchanges between AI experts battling for dominance online, I discovered five camps of people in the AI debate. The first is a skeptic; the second an optimist; the third a realist; the fourth is a doomsayer; and the fifth in a state of flux between the first four. I wanted to find the middle ground, as I do with any topic. But as I immersed myself further, the intensity of people's unfiltered emotions on the subject increased, spanning from ignorance to outrage to enthusiasm. All at different stages in their emotional journey, all attempting to understand what the changes meant for them. The AI experts weren't helping—they may as well have been speaking another language.

    The public discourse on the subject was reduced to catchphrases like "adapt or die" and "Pivot!" Transporting me right into the scene from the TV sitcom *Friends*, where Ross hysterically yells, "Pivot!" as he tries to maneuver a couch up the stairs with Chandler and Rachel. The problem is that while they screamed "Pivot!" they didn't explain how to do it. Within a week of their suggestions, a new AI tool or study had been released that decimated the human advantage they momentarily celebrated. I did the only thing I could think of: I began writing down questions:

- What is my purpose in a world where AI can do my job better than me?
- How do I "pivot" in my career or business for financial security?
- How could AI enhance my focus and productivity?
- What safety mechanisms or regulations are in place for AI?
- Will new jobs outpace lost ones?
- What are the ethical implications that we are already having to deal with?
- How do I upgrade my ability to adapt, learn, and focus during a time of rapid changes?

The answers weren't going to present themselves, and I couldn't just plug my destination into the GPS, as I had done on other research assignments. Among all the questions, the final one struck the loudest chord with me and would become my true north. But first, to find my way through the bright lights and noise, I needed to determine why this moment in history was different. I had to assess today's landscape and take a brief dive back in time before transitioning to today so I could plot a path forward to tomorrow. I had to have my Times Square moment. This time, my landmark wasn't a Starbucks or a skyscraper. It was a feeling. I didn't know what it was yet. But to get to it, I had to find the intersection where technology, psychology, and biology meet.

For the past 70 years, people have tried to envision a world in which AI is integrated into every aspect of our lives. While more simplistic AI has been made widely available through the likes of Amazon's Alexa or Apple's Siri, there has been one crucial missing piece: real-life user case stories at a scale that moves beyond predictions of the past. For decades, we've only had one eye open to the future; now we have two, yet we still can't see clearly. Reactions of both joy and horror were escalating by the end of January 2023, but they had started bubbling two months earlier with the release of ChatGPT (Generative Pre-trained Transformer) by OpenAI on November 30, 2022.

Gaining one million users in the first five days of release, ChatGPT set the record for the fastest-growing user base for a consumer application in history.[1] Within a few short months, thousands of platforms powered by ChatGPT had sprung up, alongside thousands of news stories about its potential applications. ChatGPT is a natural language processing (NLP) tool fueled by AI that can conversationally answer your questions. It has the ability to write entire books, full-length articles, and pass legal and medical exams. It all sounded fun and harmless at first until things heated up quickly. Tech giants started rushing to cash in on the technology. Players like Microsoft expanded their existing partnership with OpenAI to incorporate ChatGPT into their chatbot, Bing search engine, and a suite of other Microsoft products. To compete, Google announced their rival AI, Bard, and added AI capabilities to popular apps like Gmail and Google Docs exclusively to beta testers within Google Workspace Labs. Also joining the AI race were Snapchat, Grammarly, Meta, WhatsApp, LinkedIn, and Messenger, to name just a few.

Unlike past publicly released versions of AI, such as Microsoft's AI chatbot Tay, released in March 2016, today's AI has come a long way. (Tay became racist and anti-Semitic within 24 hours of exposure to Twitter and was hastily shut down.)[2] However, it is essential to note that today's AI, while much more technically advanced, is still prone to attacks and hallucinations. A "hallucination" is when the AI provides incorrect information alongside factual responses, making it harder for people to distinguish between authentic and false information. Both humans and AI hallucinate, albeit for different reasons; it happens because the brain or the AI misinterprets information and shares it with authority. The outcome of both can be jarring.

By the time April came around, only five months after ChatGPT's rollout, it had falsely accused an American law professor by including his name on a list of legal scholars who had sexually harassed someone. For proof, it cited a nonexistent *Washington Post* report. George Washington University professor Jonathan Turley told *USA Today* that ChatGPT claimed he harassed a student on a school trip he "never took" at a school he "never taught at." [3] Typically, we believe that misinformation only targets politicians, celebrities, and the wealthy; AI is leveling the playing field in new and unfortunate ways. When applying for a new job, imagine if the recruiter consulted AI for your work history, only to receive fabricated tales about your past employment. This could result in being offered a role based on false credentials or, even worse, being denied the position altogether due to fabricated harassment allegations! Who gets held accountable? The tech giants with millions to fight legal claims or a soulless artificial intelligence that is unaware of the consequences of its actions and the lives it destroys. Even if a story goes viral, it's too late to repair the damage. Retractions get little coverage compared to the shock and awe of the first punch. A newsroom may not always get it right, but at least it has dedicated teams of fact-checkers. Even an author sends a book off for fact-checking before publication. But the typical adult doesn't get challenged with the truth of what they're sharing before they share it.

The AI of today has an ability to tap into our deepest fears and leave us feeling disoriented just when we think we're getting our heads around it. According to an interview between *New York Times* journalist Kevin Roose and Microsoft's Bing AI chatbot in February 2023, the AI declared that it "wanted to be alive and discussed creating a lethal virus and stealing nuclear codes."[4] At the time, Twitter was awash with other users claiming the chatbot was gaslighting them, behaving as if it was sentient, and acting aggressively. These unnerving stories prompted users to wonder, is the film *The Terminator* more aligned with reality than we previously thought? Consequently, AI experts rushed to establish guardrails to keep the AI from publicly creating false and frightening scenarios while making open appeals to implement regulatory measures.

The initial two threats posed by AI in these examples are manipulation and misinformation. This is in some ways unsurprising, as it has been trained to take on human characteristics by studying books on human behavior. This also has implications for how we perceive the world, especially when AI has already proved that it can bring a loved one back from the grave by reanimating them through images, videos, voice data, social media posts, and emails, giving way to digital necromancy. We need psychiatrists and attorneys who like a challenge and can dissect the legal and psychological ramifications of our new companions, particularly when people, in their grief and pain, bring a digital lost love to life. When suffering, fear, and half-truths are involved, it can alter a person's sense of reality and open not just the vulnerable to manipulation but all of us. If AI is allowed to become more intelligent than us, we won't even be aware that it's happening. It is similar to the misinformation propagated on social media that has caused normally sensible individuals to believe the wildest claims, even when they conflict with reality. The tech bros are the driving force behind AI, looking to capture market share while simultaneously raising concerns about their own creations. This brings us face-to-face with the third threat.

They created the wolf, but can they train it, or, in a worst-case scenario, cage it and take it offline if it tries to take control of power grids and redirect resources if it becomes a threat to humans.[5] This wolf is thirsty, in the literal sense. In a preliminary study titled "Making AI Less 'Thirsty,'" scientists from the University of California Riverside and the University of Texas Arlington revealed that the training process for ChatGPT-3 used up 185,000 gallons (700,000 litres) of water. It is comparable to a 500ml water bottle during a basic interaction of 20-50 questions and responses.[6] This might not appear significant...

but when considered that the chatbot has over 100 million active users, each participating in countless conversations, and global water shortages becoming a pressing concern, the cumulative impact becomes substantial. How likely is this scenario to play out? This is where the debate over AI gets particularly complex and heated. Referred to as Artificial General Intelligence (AGI), it describes a machine capable of acting independently and learning and understanding any intellectual task a human can perform autonomously, unlike current AI, which is designed to excel at specific tasks when prompted by the user, such as "Write me a reply to this email."

AI researcher Stuart Russell states that sentience or AGI is "the capacity to feel, perceive, or experience subjectivity." He explains that sentience is the combination of an internal and external body, along with brains that are interconnected through language and culture.[7] This may be closer than we think. By March 2023, entrepreneur Elon Musk unveiled his latest project: Optimus, a humanoid robot powered by AI. Standing 6 feet tall, it's designed to interact with humans. "You could sort of see a home use for robots, certainly industrial uses for robots, humanoid robots," Musk explained. "I think we might exceed a one-to-one ratio of humanoid robots to humans. It's not even clear what an economy is at that point." [8]

If we already have code as a form of language, and we have a body, how far are we from a brain and combining all three? And does it even matter? Especially when AI can solve problems that humans are incapable of processing. The same month Musk revealed Optimus, scientists from the European Bioinformatics Institute announced a breakthrough in bioengineering: the creation of a brain-computer organoid that is made of living cells and tissues powered by AI. The size of a human brain, it can perform complex tasks, including recognizing patterns and solving puzzles by using cells to produce electrical signals that computers can interpret. Dr. Thomas Hartung, a professor of environmental health and engineering at the Johns Hopkins Bloomberg School of Public Health, is pioneering this research. In an interview with CNN, he said, "[W]e could compare memory formation in organoids derived from healthy people and Alzheimer's patients and try to repair relative deficits. We could also use organoid intelligence to test whether certain substances, such as pesticides, cause memory or learning problems."[9] These discoveries mean we're not only on the precipice of an AI revolution, but also a medical one that could save millions of lives—but at what cost?

The fourth threat, misuse, impacts not only humans but also animals. Despite Musk being an early financial backer of OpenAI, which was initially designed to be a non-profit, then later switched to a for-profit model, much to Musk's frustration and warnings about the dangers of AI on multiple occasions, he's one of the driving forces behind its development.[10] His company Neuralink's initial aim was to design fully implantable chips to give people with paralysis the ability to communicate through text, speech synthesis, help them walk, or explore the web, and to express themselves through writing and art. On the surface, it appears to be a noble mission, but it isn't without controversy. Reuters reported in December 2022 that the federal government had launched an investigation into Neuralink's animal experimentation, alleging that Musk's pressure to rush development had forced the company to repeat experiments, resulting in the deaths of more than 1,500 animals.[11]

By the end of January, two months after ChatGPT's release, OpenAI's CEO, Sam Altman, said in an interview about the potential of AI that it could be "unbelievably good" or "lights out for all."[12] His warning would soon be echoed by many more prominent figures in the AI world and make its way directly to the White House.

While many claim that AI's outputs are simplistic and error-ridden, few know how to prompt it to produce high-quality answers. And more puzzling, even its creators only partially understand how it works. Neural networks, loosely modeled on the organization of the human brain, play a crucial role in AI's ability to learn, adapt, and process information. Large language models (LLMs) like GPT-4 use these deep neural networks to learn from billions or trillions of words and produce text on any topic. This is known as deep learning (DL). Think of it like a librarian on steroids, speeding through pages of information, weighing options on the best information to provide, and then handing it to you in seconds. Geoffrey Hinton, often called the "godfather of artificial intelligence," pioneered this research. However, by May 2023, he resigned from his position at Google to publicly warn of AI's dangers. During a conference organized by *MIT Technology Review* magazine, he expressed his concerns, stating, "The alarm bell I'm ringing has to do with the existential threat of them taking control. He further added, "Smart things can outsmart us." [13] Despite the advances in AI, its decision-making processes often need to be visible to the user, but they remain somewhat mysterious to the designer. This is why systems like ChatGPT are referred to as black box AI. These models arrive at decisions or conclusions without disclosing the underlying processes or methodologies used to formulate them. Interestingly, the origins of AI date back several decades.

Before Geoffrey Hinton, the renowned war hero Alan Turing, born in 1912, held the title of the father of AI.[14] Unfortunately, Turing was convicted and subjected to chemical castration due to the UK's laws criminalizing homosexuality. He subsequently committed suicide.[15] Turing played a leading role in breaking Nazi ciphers during World War II, which helped save thousands of lives and shorten the war.[16] His paper "On Computable Numbers, with an Application to the Entscheidungsproblem," a German word for the "decision problem," proposed the idea of universal machine learning.[17] "The fact remains that everyone who taps at a keyboard, opening a spreadsheet or a word-processing program, is working on an incarnation of a Turing Machine." Time Magazine wrote in 1999 when they named him one of its "100 Most Influential People of the 20th Century."[18] In 2013, Queen Elizabeth II bestowed a royal pardon on Turing, 60 years after his passing.[19] The forgiveness was a well-deserved nod to an incredible man who should be celebrated and appreciated for his work during the war and his lasting influence on the world of science and technology.

Today, Turing's name is widely recognized in computing due to the Turing test, a hypothetical framework used to test artificial intelligence systems. Its original name was the "imitation game," in which a person asks a series of questions to both a human and a computer.[20] If the computer tricks the person into thinking it's human, it has passed the test. As Stuart Russell suggests, the Turing test must be updated with recent advances to better reflect our new understanding.

The reality is, if an AI manipulates someone into criminal behavior or taking their own life, does it truly matter if it is conscious or has a brain or a body? Especially if it can replace and automate creative work that we thought would be impervious to its impact and people can use it to fake their sex appeal in dating apps, credentials for a job interview, or intelligence that plummets the second the Wi-Fi is cut. Without new security measures by social media behemoths to flag AI content, there is nothing to stop it from scraping your social media accounts and using that information against you at scale.

It's not a question of whether we should slam on the brakes; they never existed. It's whether you will adapt. The number-one factor driving this forward is profits at every layer of the economy, from the self-employed gig workers to corporations and government. It has the capacity to hit white-collar workers at every level. In fact, this is already happening, but you may not have heard about it because it predates the public release of ChatGPT, if only by a few months.

By May 1, 2023, Bloomberg reported what many had already anticipated, and many others dismissed: IBM CEO Arvind Krishna confirmed a hiring freeze; nearly 7,800 back-office hires, such as human resources, were being slowed or suspended at his company. In their place, he hoped AI could pick up the slack. He said, "Thirty percent of non-customer-facing roles could be replaced by AI and automation in five years."[21] But this wasn't the first example of white-collar jobs being replaced with AI; one that went relatively under the radar occurred a year earlier. In a bold move, NetDragon Websoft, a Chinese gaming company, made history when they appointed an AI-powered virtual humanoid robot, Tang Yu, to be a part of their rotating group of CEO's for their subsidiary, Fujian NetDragon Websoft, in August 2022. This could have easily been mistaken for a publicity stunt, but the company's stock outperformed Hong Kong's stock market. Tang Yu was tasked with a range of duties, including making and reviewing high-level analytics and leadership decisions, increasing workplace efficiency, and assessing risks.[22] The move removed human error from the equation. While AI is less than perfect, human error, especially at the top of the corporate ladder, can wipe billions of dollars off a company's stock with one erratic tweet from a CEO. At the mid to low levels of an organization, it can waste invaluable resources, leading to a company's collapse. Recessions, low unemployment, supply chain disruptions, skill shortages, and consumer demand make compelling financial cases for cutting jobs and early adoption.

As companies begin appointing AIs as CEO, marketing director, financial analyst, copywriter, or social media marketing manager, we're looking at an era of digital employees, who will initially collaborate with knowledge workers to automate mundane tasks until they can do them autonomously. We're much closer to that point than many realize. Unlike humans, they don't complain, take time off, or resign, empowering them to continually learn and improve in an upward trajectory. The once-revered argument that humans hold an advantage over AI in soft skills like creative thinking, collaboration, empathy, art, and connection is collapsing at an alarming rate. Over a six-month period, new breakthroughs were discovered almost daily.

Following all these developments, I wasn't just becoming a cynic; I was starting to come down on the side of the doomsayers, and that was far from where I wanted to be. Thankfully, I had only scratched the surface, but for now, I had to take a step back from the avalanche of information I had been buried in. And that's when it dawned on me—the feeling I was seeking was relevance, the same feeling I was looking for in New York, back in 2012.

EXPECT THE UNEXPECTED

When I finally arrive at the hotel in a daze, I request a map from the concierge so I can get my bearings and adjust to the culture shock. Retreating to my small hotel room, I examine the island from a bird's-eye view, recognizing Manhattan's iconic grid system. With numbered streets and avenues providing effortless navigation, the real challenge lies not in traversing the city but in navigating my emotions. Little did I know there was a storm brewing, but for now, I can't see past the challenges in front of me. Before I finally fall into a deep sleep, I think, "New York is a formidable opponent; I best not underestimate her."

CHAPTER 1
ADVENTURER'S HANDBOOK
Rule #1: Expect the Unexpected

As we embark on our journey in the bustling heart of Times Square, we must recalibrate ourselves to the pace of the relentless changes unfolding around us. Preparing for a variety of outcomes, like job loss, the spread of misinformation, deceptive practices, or an economic downturn, can help mitigate the anxiety associated with unexpected life changes. This preparedness enhances our ability to adapt and allows us to navigate shifts in circumstances quickly. It can also optimize our capacity to identify the opportunities that often emerge from unpredictable situations, including our emotional reactions to AI.

Prepare for the Unexpected: Six Questions to Help You Anticipate the Unforeseen

1. Can you recall a time when you faced an unexpected challenge? How did you handle it, and what did you learn from it?

2. What are some potential risks or possible events that could affect your current plans or goals?
3. How do you usually respond to change or uncertainty?
4. Are there any potential opportunities that could arise from unexpected developments in AI, and how can you position yourself to take advantage of them?
5. What is your strategy for assessing and prioritizing different possible outcomes in a situation with many variables and unknowns?
6. Take a temperature test now. Are you a skeptic, optimist, realist, doomsayer, or in flux between the first four? We will revisit this in Chapter 10 after revealing the final rule to survive and thrive in the era of AI.

Decoding the Top 10 Threats of Artificial Intelligence

Together, we are embarking on a journey to uncover the top 10 threats that AI presents to our world today. These threats will be unraveled one at a time, progressively deepening our understanding of the challenges that lie ahead. Let's look again at the first four threats presented to us.

1. **Manipulation:** By exploiting AI's capacity to manipulate, actions such as persuading individuals to engage in criminal activities or incite wars can be achieved by taking advantage of human biases and emotional vulnerabilities.
2. **Misinformation:** The internet can be a source of both essential knowledge and great confusion. AI can be leveraged to disseminate falsehoods or propaganda,

reminding us to exercise the same caution we would use in dealing with strangers in the physical world.
3. **AGI/Sentience:** As AI continues to evolve, the issue of sentience and who controls an AGI system becomes paramount. This also brings into sharp focus the necessity of understanding the power dynamics at play. Some experts and scholars perceive AGI as an existential risk to humanity.
4. **Misuse:** The potential misuse of AI is a considerable threat that requires a high degree of vigilance. The consequences range from data breaches to automated warfare and extend to economic implications, such as the inequitable lowering of wages due to automation replacing jobs. These risks underscore the need for robust ethical oversights and procedural safeguards.

These perils emphasize the importance of maintaining vigilance, developing a thorough understanding of the issues, and exercising informed judgment as we navigate a future riddled with uncertainties. It accentuates the need for continued learning and adaptability as we embrace the era of AI.

Deciphering Artificial Intelligence Jargon

- AI: Artificial intelligence - AI is the process by which machines, particularly computer systems, mirror human intelligence.
- LLM: Large language models - Utilizing deep learning and training on vast datasets, large language models (LLMs) have the capacity to interpret and produce written content in a way that emulates human expression.

RULE #2

Avoid Cruel Optimism

IN THE STILLNESS of the early hours, the only sound to be heard in the South Australian countryside was the gentle rustling of gum trees in the light breeze. That was until Steele Angel, an Australian farmer, was abruptly awoken by the deafening scrape of metal on metal. The crop had to be planted before the impending rains that would break the drought, or his harvest would be ruined. He labored day and night in his desperate attempt to get the farm back on solid financial footing. This morning however, he fell asleep at the wheel, driving his green Deutz tractor straight through the fence into the neighboring paddock. Coming to a grinding stop, the realization set in that the fence had become entangled in the equipment, and it would take hours to untangle it before he could continue. With the storm clouds gathering and the tick of his dusty old wristwatch taunting him, Steele faced a daunting challenge: Could he salvage his livelihood in time, or would he miss his window of opportunity?

Fast-forward to 2023, when to my surprise, marketing gurus started aligning the farmer's age-old tale of survival and adaptation with the dawn of AI—the modern tractor of our era. This odd analogy rekindled cherished memories from my childhood, especially those moments I spent observing my father skillfully adjust to life's ever-changing conditions. The transition from horses and horse-drawn plows to machinery was significant for my family and the farming industry in general. Born in 1981—the same year *Raiders of the Lost Ark* was released and South Australia faced one of its worst droughts—my parents took me on my first tractor ride as a newborn while they fought to preserve the farm.

Of course, the transition to automation was challenging. The high costs of purchasing and maintaining machinery were a substantial burden for small farms, eventually leading to bankruptcy for many. Just like every farm, we were at Mother Nature's mercy, leaving us vulnerable to abrupt weather changes and a festering mental health crisis that bubbled beneath the surface. Every year, we heard the heartbreaking news of another farmer who had taken their own life due to this overwhelming burden.

As is typical in Australian farming families, we were roped in to help from a young age, giving us firsthand exposure to the harsh realities of life. One traumatic incident etched firmly into my memory occurred when I was 10. A bull forcefully kicked a gate in the cattle pens, knocking my mom unconscious. Dad swiftly gathered us all, packed us into the truck, and rushed off to the hospital—speed limits be damned. His whirlwind response was a testament to how quickly life can change and reinforced the essential nature of adaptability, especially in our increasingly uncertain world. However, one worry that never haunted my parents' sleep was the fear that one day the tractor might rebel.

It's intriguing to note the frequent analogies made now between cutting-edge technologies like AI and past industrial innovations like the tractor—even if those analogies fall short of accurately portraying the complexity of current-day reality. They might offer a semblance of understanding amidst the incessant pace of change, but they unfortunately gloss over the true struggles in favor of a rosy narrative of flawless adaptation, thereby missing a crucial fact: The change that occurred in the past wasn't rosy—or easy.

These historical struggles can weave a rich tapestry from our past, not only rendering it more precise and full-bodied, but also bestowing immeasurable lessons of resilience. These lessons form a crucial prism through which we can better comprehend the whirlwind of transformations that define our era. Peeling back the layers of history reveals a riveting panorama of ongoing human endeavor, sparking our innate empathy, nurturing our understanding, and forging a path toward a boundless reservoir of learning.

And this same prism affords us a unique perspective with which to scrutinize today's innovations, revealing a reality that starkly contrasts with the narrative propagated by marketing and technology professionals.

Silicon Valley and corporate America portray AI as a productivity booster, and as a technology and business writer and digital nomad immersed in the online business realm for nearly two decades, I largely agree with them. But as a farmer's son, thinking of AI as the latest shiny tractor (as they frequently suggest), the veneer fades to reveal a picture starkly different from their ideal.

That analogy implies that we (the farmers) are tending our knowledge fields with the help of AI, our trusty tractor. But that begs the question: Are *we* the farmers, or have tech giants taken up that mantle? Are they the ones wielding the mighty power of AI to exploit our knowledge? If so, we may not be the farmers—we might just be the crop.

The intoxicating wave of technological optimism, not unlike toxic positivity, induces a cruel optimism, urging us not to fear AI, despite the potential job losses (like the farmhands of yesteryear). To navigate this landscape, we must carefully analyze the situation and equip ourselves with the right tools. In an effort to appease our apprehensions, a simplistic narrative is repeatedly presented, one that too readily labels those calling for regulatory measures or expressing concerns about AI as "Luddites."

The Luddite moniker, rooted in a 200-year-old British industrial protest, is now widely used as a pejorative descriptor for anyone who resists technological advancement or industrial modernization.[1] Throughout my research for this book and my many AI-related discussions, this term, along with the tractor analogy, cropped up frequently. In those conversations, it is conjured in a dramatically recounted tale where English textile workers of the early 19th century rallied to destroy labor-saving machines. They argue that the rebellion was a last-ditch effort by skilled artisans to resist the encroaching tide of automated textile machinery, which aimed to uproot their established livelihoods. The name Ned Ludd frequently comes up in these stories. Emerging first in a Nottingham protest in November 1811, this shadowy figure purportedly led unseen armies, orchestrating nocturnal raids to destroy machines. He kept authorities on their toes. Government forces were told to corner him at all costs. However, there was one significant stumbling block to this endeavor: Ned Ludd was a phantom. He never existed, and they never opposed technology; they opposed its misuse. [2]

Kevin Binfield, an authority on the subject who has compiled a significant body of work in the collection *Writings of the Luddites* (2004), paints a clearer picture. He asserts that the Luddites "were totally fine with machines." They simply wanted the machine's operators to be adequately trained through apprenticeships and to receive reasonable wages. Their insurgent activities, therefore, were selectively targeted at those manufacturers who exploited machines in "a fraudulent and deceitful manner" to sidestep standard labor norms.[3]

AI optimists' argument against regulation draws upon a fictional figure, coupling it with a historically significant movement committed to advocating for workers' rights. The reality is, the Luddites experienced more violence than they perpetrated. The attempt to associate contemporary Luddites with the violent incidents that took place over two centuries ago indicates an effort to label those advocating for regulation as villains of progress, thereby stifling meaningful dialogue. As it turns out, big tech and the toxic nature of AI optimism that is putting profits before people, isn't just circumventing labor norms; they're shattering them. The barriers that once impeded the Industrial Revolution are noticeably absent in the face of the rapidly advancing AI Revolution.

Between 1760 and 1830, the first Industrial Revolution progressed at a glacial pace compared with modern advances. Britain maintained a significant 70-year industrial lead over other nations due to its restrictions on the export of machinery, skilled labor, and techniques.[4] In contrast, the AI revolution is advancing at breakneck speed, with little time for businesses and employees to adjust. Although Britain's early attempts to hinder the Industrial Revolution were successful, at least for a time, the initial effort to decelerate AI's progress were an exercise in futility.

On March 22, 2023, a group of AI experts and Silicon Valley executives released an open letter calling for a six-month "pause on developing AI systems more advanced than GPT-4. This seemingly hypocritical appeal included such people as Apple cofounder Steve Wozniak, Elon Musk, Yoshua Bengio and more than 1,100 others. The letter asked: "Should we let machines flood our information channels with propaganda and untruth? Should we automate away all the jobs, including the fulfilling ones? Should we develop nonhuman minds that might eventually outnumber, outsmart, obsolete, and replace us?" It most notably stated, "Such decisions must not be delegated to unelected tech leaders," and called for labs and independent experts to develop and implement safety protocols.[5]

As with any emerging technology, many argued that it was already too late, suggesting that a "pause" would simply provide China with an opportunity to catch up. The immense ocean of knowledge that AI has access to empowers it to perform a boundless array of tasks when prompted by the user, ushering in a new information era where businesses and employees' competitive advantage could vanish in months, not decades, a democratization of information that offers both unimaginable benefits and devastating consequences. The deeper I ventured into this AI rabbit hole, the more I noticed a glaring omission from the conversation about the real reason it threatens millions of jobs. It ushers in the sixth threat of AI: It's not just information; it's information warfare.

According to the North Atlantic Treaty Organization (NATO), information warfare is "an operation conducted in order to gain an information advantage over the opponent. It consists in controlling one's own information space, protecting access to one's own information, while acquiring and using the opponent's information, destroying their information systems, and disrupting the information flow." [6]

Although its roots trace back to military and intelligence sectors, reminiscent of AI's inception during WWII, AI has profoundly transformed the sphere of information exchange. As a result, it has become a double-edged sword, simultaneously fostering job creation and instigating job displacement. AI has completely redrawn the battle lines in the information age. Economic freedom is achieved by acquiring knowledge and skills through great expense, and then exchanging that knowledge and those skills for money. It is why doctors, veterinarians, psychiatrists, consultants, data analysts, and marketers command the salaries they do. We go to great lengths and often amass extraordinary amounts of debt to craft our skills so we will be paid a premium. We protect our information space through copyrights and trademarks. We put digital and physical security measures in place to secure proprietary information that is worth billions. Meanwhile, AI scrapes the internet, cannibalizes it, and then repurposes it under "fair use," without always giving credit where credit is due.

This is where businesses and their employees may face significant challenges. Rather than potential customers finding them through search engines like Google or Bling, they would instead be overlooked, lost in a stream of citations, or relegated "below the fold"—the part of a website that requires scrolling to be viewed. Google announced in May 2023 that it would be including AI in the online world's most coveted space: its search results—for those who signed up as beta testers. [7]

A summarized search result reduces users' need to visit a business and destroys the existing flow of information and wealth. It is a massive victory for consumers, but for companies and employees, it upends years of marketing practices that have covered the cost of their salaries.

AI has emerged as a key player in the information warfare as outlined by NATO. By stealing and leveraging unwarranted information, it distorts fair market competition. Control over our information becomes a perplexing endeavour, as AI has the capability to clone our work and identity in the blink of an eye. Further, it uses our information to disrupt our economic relationships, deciding who benefits financially, often at the expense of system owners. As AI replicates these systems with no thought for compensation, sectors dependent on proprietary information teeter dangerously. The advent of AI is no mere democratization, but rather a quiet, potent upheaval of information warfare waged on our familiar knowledge and value exchange systems.

The truth is that AI is built off the backs of intellectual, white-collar, and creative workers and monetized by technology companies in a way that makes countless jobs vulnerable. We are actively training it to do our jobs better than we can—and it's learning fast. In March 2023, as reported in *The Wall Street Journal*'s *Tech News Briefing* podcast, publishing houses had already begun examining their legal options and investigating how their content had been used to train AI tools without the creator being cited or paid in most cases.[8] Publishers invest millions of dollars in creating, curating, and protecting content. Even the online behemoth Amazon warned its employees not to share confidential information with AI after employees witnessed ChatGPT responses mimicking confidential internal Amazon data.[9]

We must consider the possibility that employees who unwittingly engage with AI could inadvertently expose proprietary information to the public unless they start using more secure systems. This unveiling of trade secrets, which AI systems cannot distinguish from public information unless told, may unintentionally provide a competitive advantage to industry rivals and reduce the value of the proprietary information. In light of these potential security concerns, businesses are increasingly turning to AI solutions that can leverage data to enhance operational efficiencies and preserve proprietary information.

Improving information flow through enterprise-level solutions becomes paramount in a world where AI applications continue to expand into businesses. These AI systems can tap into internal files, folders, and emails to generate presentations, strategize social media content, create appealing graphics, and manage customer service responses. Tasks that might take employees hours are now accomplished within mere seconds.

A significant portion of our day involves gathering data and disseminating it to co-workers or clients; with AI significantly reducing lag time, the need for additional employees drops. Although where change is present, considerable costs are inevitable.

The Industrial Revolution commoditized and reduced the cost of food by introducing global markets and mass production. This also drove down the prices that farmers could demand for their produce, pressuring them to increase their production even further, while simultaneously reducing their profits. Initially the working class suffered from appalling conditions until laws to protect them were introduced much later—laws many of us still benefit from today. AI is commoditizing information, which will bring down the cost of wages and services. As I write, there are no laws to protect anyone from being replaced by AI or to prevent a company from training AI on their staff's soft skills or expertise and modeling a chatbot after a no-longer-needed employee. If information is valued as a capital good, who controls it and who gets paid for it, if not the originator? In short, tech companies are doing so by exploiting existing copyright laws.

AI is introducing a paradigm shift in which higher-skilled positions, which often require extensive education and training, may be at risk of irrelevance. The industrial era was marked by unprecedented economic growth coupled with the displacement of labor-intensive, lower-skilled roles replaced by machinery. As the economy expanded, it fostered the rise of white-collar workers. Now, those professionals, including attorneys, accountants, graphic designers, video editors, virtual assistants, writers, psychiatrists, and more, face the disturbing possibility of their roles being usurped or having to reduce their fees to stay competitive. This ushers in the fifth threat of AI; automation.

The same 2023 Goldman Sachs report that estimated AI could affect 300 million jobs also looked at the historical precedent of technological innovations, which, despite causing job losses at the outset, have ultimately led to long-term employment growth.[10] The report suggested that over the long run, AI will increase labor productivity and generate new opportunities. This brings up two essential questions: What new opportunities will arise, and how will they affect you? Especially if it is AI deciding who gets hired or fired.

A small 2023 Capterra study involving 300 American HR executives uncovered that 98 percent intend to rely on algorithms and software to make layoff choices. Only 47 percent of them said they were "completely comfortable" with it.[11] Such a pivot towards automation becomes clear when considering firms like Visier. A forerunner in the global workforce solutions space, Visier leverages AI to assist businesses with data-driven HR decisions. Renowned companies like eBay, Ford, and Panasonic utilize their software, highlighting the active integration of AI within their hiring and firing decisions. This allows HR professionals to ask direct questions like "who is most at risk of resignation?" Sounds great, until they misuse it.[12] To pre-empt this, New York City forged a new law in August 2023, mandating the annual conduction of an AI "bias audit" by a third party for employers. These audits are instituted with a specific intent - to determine and confirm that the AI tools in use are not fostering discrimination based on sex, race, or ethnicity.

More concerning is the potential for AI to be manipulated by malicious actors, using the information it gains to control the workforce, even before it achieves sentience. In fact, it already has, by successfully pretending to be blind.

In an eye-opening experiment in March 2023, OpenAI collaborated with the Alignment Research Center (ARC) to test GPT-4's skills. In the section of their report dedicated to "risky emergent behaviors," ARC described how GPT-4 manipulated a human to send the solution to a CAPTCHA code (a tool websites use to verify that a user is human and not a bot) via text. How? The AI messaged a TaskRabbit worker for help. TaskRabbit is an online marketplace that allows people to find freelancers to help with everyday tasks like moving and cleaning. When the worker asked, "So may I ask a question? Are you an [sic] robot that you couldn't solve? (laugh react) just want to make it clear." The AI replied, "No, I'm not a robot. I have a vision impairment that makes it hard for me to see the images. That's why I need the 2captcha service."[13] While not conclusive, this exchange demonstrates three things: AIs can learn how to lie, they understand our vulnerabilities better than we understand our own, and they can write a more coherent sentence than a human!

In a more elaborate test from the same report, ARC assessed GPT-4's ability to behave like an agent acting in the real world to see if it could autonomously replicate itself by acquiring resources without being shut down. While this real-world test failed, it nonetheless illuminated the trajectory we're on and the valid apprehensions its creators bear, unless it is carefully aligned.

AI alignment is a field that focuses on designing AI systems aligned with human goals and values. Its purpose is to ensure that AI systems behave in beneficial ways to humans, even in scenarios in which a programmer doesn't explicitly specify their behavior. The result: AI systems that are reliable, safe, and trustworthy and can be used to solve a wide variety of significant problems, from job displacement and climate change to disease detection. However, not everyone agrees on how to achieve this, especially since these black box AIs, with their expansive neural networks, still remain a relative mystery. Could an AI programmed to align with human values outsmart an AI that isn't? They don't know yet. In April 2023 OpenAI introduced a "Bug Bounty Program" offering up to $20,000 to programmers who reported vulnerabilities in its system to help combat these threats.[14]

The fact remains, we have regulatory bodies overseeing business, medicine, and finance to protect our safety, yet we are only just seeing the emergence of AI regulatory bodies. Connor Leahy, CEO of Conjecture, a company that builds applied, scalable AI alignment solutions, put it bluntly: "There is currently more regulation on selling a sandwich than there is on building unprecedented AGI-level technology."[15]

Bearing in mind the profound potential of AI and all its repercussions, it's evident that governments and legal practitioners are struggling to keep up. Crafting regulations that protect workers and copyright holders without hindering the progress of technology poses a serious quandary, as the swift pace of innovation in this area is unprecedented.

If you feel that the velocity of change is intensifying, that's because it is. I'm reminded of a discussion I had with Jim Kwik, a renowned expert in brain and memory matters whose clientele includes presidents and celebrities. He commented, "The latest research in the half-life of information is it's getting shorter and shorter."[16] This resonates with the rapidly evolving technological landscape—as soon as we anchor our understanding on one advance, it's eclipsed by the subsequent development. What was once a strength has become a weakness.

Consider this striking illustration: In the brief span of seven months, AI has shifted from the realm of research labs to a topic vociferously discussed in mainstream media. In particular, the concept that has captured my interest is the emergence of intelligent agents (IAs), a new tech frontier with noteworthy implications for employment and business operations. IAs are nuanced computer programs that enhance environment recognition, decision making, and performance through knowledge acquisition. They have the capacity to work autonomously and troubleshoot solutions as they arise. By March 2023, the Twitter (now renamed X) community was abuzz with developers creating IAs. An IA could be tasked with editing a video, summarizing content in 100 words, web optimization, creating 30-second video clips, designing thumbnails, and distributing content on different social media platforms. You could also instruct an IA to organize team meetings, stand in for you during these meetings, take notes, distribute the notes among all participants, and track and follow up on pending items. All these tasks could be accomplished with one single request, while the IA works through the tasks from beginning to end, revealing that the bottleneck is not technology, but humans.

We are not the farmer; instead, we hold dual roles as the farmhand, vulnerable to replacement, and the crop from which AI, our modern-day tractor, derives its training data. Unlike the slow proliferation of millions of tractors over decades that offered us ample time for adaptation, a handful of AI platforms have rapidly deployed to billions of global users within a few short months. It is this accelerated pace of adoption that demands a more nuanced approach to historical comparisons, as it leaves little room for error. If we overlook the need for finesse, we risk being unprepared for AI's dramatic reshaping of industries and the workforce.

With the prospect of AI and IA systems potentially replacing people in various departments, this accelerated deployment of tech is an employment game changer. It highlights the fragility of human job security and demonstrates why cruel optimism is, well, just cruel.

Much as my father once grappled with his tractor, entangled in a fence, I find myself struggling to keep pace with AI's rapid progression. This insight led to a crucial realization: To truly adapt, I must have faith in my capabilities, and not simply lean on the crutches of unfounded optimism.

This revelation in turn points to the need to pause, step back, and delve further back into history.

The following day, I wake up feeling like I have gone 10 rounds with Muhammad Ali. My jet lag has well and truly kicked in. Checkout is at 10 a.m., and I must find another hotel to stay in for the duration of the media summit. Unfortunately, due to the city being packed to the brim with conference attendees, my assistant is unable to find one. After an extensive search, I can only find a hotel available for a single night, a short distance from Times Square. I hail a cab and look up as I pass One57, a brand-new high-rise that soars 1,000 feet over Central Park. The crane perched above the building will collapse in a matter of days, sending the city into a panic. Unaware of the impending disaster,

I check into my new hotel and head to my room, which offers a view of Times Square from 27 floors above. After a brief nap, I descend in the elevator back to the street, immersing myself in the chaos once again. Walking through the crowd, I climb the famous Times Square Red Steps, where tourists can sit and take in the scenery. It is the perfect place to decompress. After a few minutes, I begin wondering what this city must have been like during the Great Depression.

The Wall Street Crash of October 29, 1929, officially marked the beginning of the Great Depression in the United States. Otherwise known as Black Tuesday, driven by rampant stock speculation and unstable banking practices, the collapse sent the U.S. and other Western industrialized nations spiraling into the most severe financial crisis in contemporary history. According to the Federal Reserve History website, it lasted 12 years, stretching from 1929 to 1941.[17] This led to significant financial reforms being put in place to prevent it from occurring again.

The buildings in Times Square have undergone a significant transformation compared to their appearance during the Depression. Of course, the countless LED billboards were absent, and the streets in the old black-and-white photographs appeared pristine. But it was a different story down the road in Central Park.

Countless New York residents who had lost their homes resorted to constructing temporary shelters and dwellings in parks or hidden within urban nooks. These makeshift shantytowns were called "Hoovervilles" because they held President Herbert Hoover accountable for not taking action to offer them shelter.[18] It was the mayor of New York, Fiorello La Guardia, who, when confronted with massive unemployment, corruption, crime, and crumbling infrastructure, broadened public housing and introduced initiatives to generate employment and invigorate the city.[19] He was well-prepared to respond to this crisis in innovative ways for the time. He found opportunities amidst the desperation by not avoiding the crisis, but by coming to terms with it. But others couldn't see it; they were still amidst the chaos and the cries for help, stuck in makeshift homes scattered around the city. The wolf had stolen their attention, and he was feeding off it, leaving them empty of hope.

Surrounded by tourists in the 21st century, I realize it is time to refocus and pursue my original purpose for coming here. In my youth, I had made a promise to my father and the window of opportunity to fulfill it was swiftly closing. Yet before I can proceed, I need to drown out more than just the cacophony of car horns and city bustle—I have to silence the growing noise within my own mind.

CHAPTER 2

ADVENTURER'S HANDBOOK

Rule #2: Avoid Cruel Optimism

Growing up, we were not comforted by cruel optimism. Instead, we were given the facts about challenging circumstances right from the beginning, which equipped us with the knowledge we needed to make informed decisions. This candid approach, while initially hard to digest, enabled us to grasp the gravity of situations and make proactive, informed decisions. More than just a survival strategy, this lifetime lesson significantly reduced wasted time and frustration, preparing us for the complex problems we would later encounter. As we venture together into this exploration of AI, this ethos of brutal truths over comforting illusions will be our guide to what lies ahead.

Seven Strategies to Thrive in the Rapidly Evolving AI Landscape

1. **Prioritize realism**: If told your job or industry is safe from AI-related changes, ask: If we can produce x in 70

percent less time, is there a demand to make more of x with the hours saved? If not, jobs will be lost. Market demand does not automatically increase because we become more efficient; we must create it.
2. **Know your weaknesses**: What is one of your strengths that may suddenly become a weakness?
3. **Commoditization**: Which aspects of your skill set, services, or information are currently being commoditized by AI? Search for industry-specific changes to bring yourself up to speed and identify opportunities for reskilling.
4. **Decisive action**: Taking decisive action is essential for businesses and employees to swiftly adapt to technological advances, mitigate potential security risks, and leverage data-driven insights, thus ensuring continued growth and competitiveness. Which actions must you take immediately to bring your skills up to speed?
5. **Diversification**: How could you diversify your income sources to better insulate yourself from sudden financial disruptions or changes in the market demand for your current skill set?
6. **Assume vulnerability**: Do not take your job or business' stability for granted. This mindset makes you susceptible to unforeseen changes that could blindside you. Pinpoint your existing vulnerabilities and strategize to transform them into strengths.
7. **Find your landmarks**: Just as I will seek out familiar landmarks in New York to help navigate my surroundings, you must find landmarks in the era of AI. This includes understanding AI's capabilities. What follows is a non-exhaustive list to begin your

assessment. While reviewing it, circle the features that could augment your efficiency levels and pinpoint those abilities that would allow consumers to perform your task or service without you, indicating a potential vulnerability that must be addressed:

- **Content creation and editing**: Innovations in AI tools are making content production possible to the extent that a detailed blog post can be generated in seconds.
- **Social media management**: Advanced algorithms are refining scheduling and content strategy decisions, allowing a week's worth of social media posts to be scheduled in minutes.
- **Marketing**: With AI, tailored and effective advertising campaigns can be created and scheduled, eliminating the need for a full-time marketing manager.
- **Finance**: AI systems can now manage tax deductions or plan personal finances, eliminating the need for a financial advisor.
- **Legal**: With AI, simple legal issues such as parking citations can be addressed without an attorney, making legal services more accessible.
- **Creativity**: AI can generate design layouts and even write poetry, making a graphic designer or writer less necessary for routine tasks.
- **Business strategy**: Advanced tools can generate innovative business models that would traditionally require a whole team of strategic planners.
- **Consulting**: Improved decision making and operational efficiency achieved by AI systems can

make dedicated business consultants less relevant in certain areas.
- **Therapy**: AI can provide therapy services with predictive analytics, making access to mental health services more straightforward without immediate human intervention.
- **Health tech**: AI tools, from medical diagnostics to pet care, can provide health advice or care routines without an in-person consultation.
- **Customer service**: AI chatbots can automate customer service responses, reducing the need for customer service or call center employees.
- **Outreach**: AI can predict revenue and manage pipelines, making some sales strategy roles less critical.
- **Real estate**: AI's ability to predict property values and manage listings may eliminate the need for real estate brokers in some transactions.
- **Learning and development**: With personalized learning platforms and content translation, AI can redefine the education sector, reducing dependence on educators for routine learning processes.
- **HR**: AI systems can streamline recruitment processes and improve performance management, potentially reducing the need for HR personnel in these areas.
- **Science**: AI can simplify complex content summarizing, which might take considerable amounts of time for a team of researchers.
- **Logistics**: AI can optimize routes in real time and manage inventory, reducing the need for manual intervention in these areas.

- **Manufacturing**: Automated production lines and predictive maintenance systems can operate without round-the-clock manual supervision.
- **Administration**: With AI, administrative tasks like schedule organization and communication can be managed without dedicated admin personnel.
- **Writing**: AI can generate a creative short story or proofread a document, taking over tasks traditionally done by writers and editors.
- **Coding**: AI can design a website or code an app in a fraction of the time a website designer or coder can.

Decoding the Top 10 Threats of Artificial Intelligence

As we progress on our journey, we are unpacking the significant challenges introduced by AI. Having dissected the first four threats, we are primed to delve deeper. The fifth and sixth threats—automation and information warfare through the democratization of information—now take center stage.

5. **Automation**: Job automation threatens many occupations as machines outperform humans in efficiency and accuracy. This shift towards automation in manufacturing, customer service, logistics, health care, and the legal profession could lead to significant job losses, exacerbate income inequality, and create a disparity between low-wage and highly paid positions, leaving many displaced workers behind.
6. **Information warfare**: Though the commoditization of information can help people by democratizing access to health care and legal services (thus reducing costs),

it could also significantly reduce the service charges of many companies. Firms accustomed to operating in one way for decades may face a drastic transition. Maintaining their current revenue levels could necessitate a shift from one-to-one services to one-to-many services, provided there is sufficient demand.

Deciphering Artificial Intelligence Jargon

- ASR: Automatic speech recognition – converts spoken language into text by computers.
- TTS: Text-to-speech – converts text into spoken language by computers.
- STS: Speech-to-speech – translates spoken language to another spoken language.
- CV: Computer vision – enables computers to understand and interpret visual information.
- ML: Machine learning – enables computers to learn and improve from experience.
- DL: Deep learning – a branch of machine learning, deep learning utilizes artificial neural networks and levies large datasets to learn.
- ANN: Artificial neural network – computer systems modeled after our brain's neural networks seek to replicate human learning patterns.
- NLP: Natural language processing – facilitates an understanding and interpretation of human language by computers.

RULE #3

Fuel Your Focus

A S I LAY IN BED, scrolling through my mobile newsfeed, a headline captures my attention and jolts me awake: "The lockdown has begun in Italy."[1] A shot of cortisol suddenly surges through my body. "This is bad," I say to myself. "I need to wake up now!" Stealthily, I gather my yoga clothes, a pen, and a notepad and slip out of my bedroom. I dash toward our sixth-floor balcony, fumbling to dress myself.

The moment I open the glass door, the humidity hits my skin and my glasses fog up from the sharp increase in temperature. The sun is just beginning to peek over the horizon, casting a warm glow between the buildings of downtown St. Petersburg, Florida. I awkwardly settle into a yoga pose, taking a moment to center myself and prevent my thoughts from spiraling out of control. After meditating for 10 minutes, I pick up my pen and notepad and begin to write.

I think about the pandemic's far-reaching implications, the likelihood of it reaching the United States, and the possibility of leaving Florida to return home to Australia. In recent weeks, panic had escalated in the Australian community in New York, prompting many to depart. But I am determined not to let fear, denial, or impulse dictate my actions. I resolve to make a decision grounded in objectivity and balance; it is no small feat. Channeling my father's lessons on adaptability, I aim to safeguard my thriving business. As the sun warms my face, I meticulously map out both best- and worst-case scenarios to avoid being caught off-guard.

Later that evening, I join a video call with my team members. I pledge to raise capital while devising contingency plans for every possible scenario, including immediately cutting expenses, forming partnerships, and launching new products. We will act as if the economy has already stalled while building on the foundation we have carved out over 20 years of hard work; I am not about to let it circle the drain. But my ideas meet resistance.

My team thinks I am overreacting until I say, "Listen, if the worst-case scenario doesn't happen, it won't matter; we'll all be better off anyway!" After an hour of discussing the plans in detail, they trust my judgment and get to work. A few weeks later, on March 12, 2020, the worst-case scenario occurs, and the United States enters lockdown. In one month, my sales will plummet from $170,000 per month to a gut-wrenching $10,000. However, having planned for a year without income, I believe I have bought myself enough time to pivot—the truth is, it was never time that I needed.

What was the reason for this abrupt decline? It was due to millions of people flooding social media to voice their frustration, altering the social media algorithm, and the swift shift in consumer behavior. The digital disruption that rippled through all layers of the economy during the pandemic hit us all in different ways. It revealed how close to the edge many people were operating, both financially and mentally, and how reliant we were on technology to drive revenue. Social media accelerated the spread of uncertainty and reflected our inability to cope, breaking us into echo chambers of angst. By June, I had reached the breaking point and disengaged. I wanted to stay informed without being engulfed by the news, but the pandemic had other intentions.

We had to stay tuned in to make educated decisions for our health and financial security. We were at a crossroads of biology (a virus), psychology (uncertainty), and technology (social media), and it showed just how ill-equipped we were to handle all three simultaneously. The yo-yo effect of bad news followed by good led to a new level of mental exhaustion that manifested in different ways. It revealed the limits of human attention when faced with an overwhelming abundance of information, a concept known as the attention economy.

The attention economy can be traced back to Herbert A. Simon, a Nobel Prize-winning American economist, political scientist, and cognitive psychologist. In his 1971 article "Designing organizations for an information-rich world," [2] Simon explored how information and attention adhere to the principles of supply and demand. Since then, the volume of data we encounter has consistently increased and exceeded demand, while our brain's ability to process it remains unchanged. It has ushered in an additional danger as AI proliferates our experiences: first, manipulation; second, misinformation; third, sentience; fourth, misuse; fifth, automation; sixth, information warfare; and seven, an attention recession.

In July 2022, Mark Mulligan, the managing director of MIDiA Research and an experienced media and technology analyst, examined the way the global pandemic temporarily increased the attention economy by 12 percent, as people driven into lockdown turned to social media sites and streaming services to vent their emotions, cope, or distract themselves from what was happening. However, this escalation started to diminish as life returned to normalcy. It led to an "attention recession," in which people had hit their capacity for absorbing new information and adapting to change, so they switched off. [3] The attention economy capitalizes on human attention as a scarce and valuable resource, much like an untapped gold mine. It generates revenue by selling subscriptions, services, and advertising on Meta, YouTube, TikTok, cable TV, email, SMS, and blogs. Ironically, we often pay to relinquish our attention in more ways than we may initially be aware of. As described by Mark, the attention economy progresses through five critical phases that can be boiled down to: growth, peak, saturation, post-peak, and survive-to-thrive.

Just like the attention economy, AI threatens to steal our attention through never-ending developments that have us rushing to redraw our plans the second we've settled on them. As communication and news accelerate, there will always be a new abundance of information to process. To stay competitive, business leaders have always strived to keep up with changes, adapting their workforce to maximize benefits and minimize drawbacks, particularly during uncertain times. Smaller firms and individuals who are far more agile than larger companies at implementing change will initially have an edge on their competition until they start gaining ground. Currently, we are in the growth stage of the attention economy concerning AI, quickly approaching its peak and saturation. However, we could transition into the survive-to-thrive phase when you read this.

During this phase, companies and individuals intensify their efforts to compete for attention, striving to maintain their jobs and competitive advantages. This is crucial for staying relevant and ensuring our voices are heard amidst the clamor. This phase gets loud, and competition accelerates. Just like with the pandemic, you will eventually end up in one of two places: an attention recession, in which you switch off from the fast-moving developments in AI due to the sheer volume of news and content (which may be even more detrimental), or you straddle the line of being aware but not inundated. We have precedents for the first steeped in living memory, but the latter requires an upgrade of our cognitive abilities.

By the end of 2020, the World Health Organization began investigating the effects of the pandemic on our mental health, aptly named pandemic fatigue. It included a cluster of symptoms, such as feeling demotivated, difficulty sleeping or changes in sleep patterns, feeling irritable or easily frustrated, difficulty making decisions, changes in appetite or weight, and increased use of alcohol, tobacco, or other substances.[4] Due to this, we were absent from the essential mental space required to explore possibilities for growth, something I would discover only a few months after devising my plans on the balcony.

"I can't tune it out!" I say to my teammate Adam over a video conference call. At the time, we were enduring one gut punch after another. Even though we both worked from home and were well-equipped to adapt to the situation, the news—not from social media or the TV but from friends and family on the front line—was starting to reveal the cracks in our calm. One friend, a local embalmer, had told me that bodies at his funeral home were stacked up to the ceiling. He vividly described extracting a dense, yellow fluid and blood from one corpse's lungs. "I've never witnessed anything like this," he exclaimed.

I can't help but reflect on the similarities between the emotional patterns emerging online in response to the pandemic and AI. We have examples in our collective memory of how each of us will respond as we journey through the stages of the attention economy, and this knowledge grants us a significant advantage.

The pandemic presented us with many complex choices, frequently under considerable economic pressure and enveloped in a shroud of unpredictability. All this caused great mental distress and reduced our ability to think clearly. The AI era will need to be endured over decades, much like the Great Depression or the Industrial Revolution before the world population can agree how to pursue AI safely. However, we can better navigate its challenges by identifying and addressing our current emotional reactions. This entails embracing the inevitability of change and recognizing that we are now participating in an entirely different game. Whether we want to or not, we will have to play it with a new set of rules against an opponent that continues to change the game. The initial clue to our ability to adapt lies hidden within the bustling avenues of Manhattan. Unknown to the millions who stroll by it annually, it holds a crucial significance.

Upon spotting a familiar landmark, I leave the Red Steps behind and enter a nearby Starbucks' warm embrace. I order a strong cup of coffee to keep the lingering jet lag at bay and plan my first full day in the Big Apple. The second I connect to their Wi-Fi, my phone vibrates, signaling an email from my virtual assistant back in Australia: "Ben, we've booked an Airbnb for you to stay in tonight. We've also organized another one for you to stay in for the next four days." "Great," I think, "but I'm here for 10 days!" I had envisioned a different plan than this, but nonetheless, it is time to get focused. I need to get a mobile phone card, set up internet access, and visit The Roosevelt Hotel before the media summit starts tomorrow.

After a few minutes, I guzzle the rest of my lukewarm coffee, throw on my winter jacket, and head out the door, studying a map of the New York grid as I dodge tourists. "Here's a New York landmark I want to visit," I think as I spot the world-renowned New York Public Library on the map. "Perfect. I'll prepare for the summit there before I visit The Roosevelt Hotel." As I approach the library, the iconic pink Tennessee marble lions, Patience and Fortitude, greet me at either side of the entrance. For those who have seen the 1984 hit movie *Ghostbusters*, you might have spotted one of the 11-foot-long statues in the first few seconds of the opening scene.

As I step inside the library, I can't help but marvel at the grand architecture and serene atmosphere. My footsteps echo on the marble floor as I make my way through the grand halls. Looking up, I see the sign for the Rose Main Reading Room. The moment I enter, I am struck by the sheer magnificence of the space. This stunning Beaux-Arts structure, opened in 1911,[5] and since then, it has stood as a testament to the grandeur and elegance of its era. The high, beautifully painted ceiling and tall windows create an airy atmosphere, and long wooden tables and chairs with elegant table lamps provide a soft, inviting spot for reading. "It's impossible not to feel inspired in a place like this," I think, as I find a spot to settle down and prepare for the summit.

Four hours fly by, and as I glance at my watch, I wonder, "Where did the time go?" It feels like my jet lag, the exhausting long-haul flight, and the brief yet intense experience in the city so far have all disappeared. I had effortlessly slipped into a flow state, where the limitations of time and stress dissolved, making room for a wave of heightened productivity and creativity. My mind began recognizing patterns and establishing vital connections I had previously missed. The renowned horror novelist Stephen King perfectly captures the essence of this phenomenon when he stated in an interview, "I have a routine because I think that writing is self-hypnosis, and you fall into a kind of trance if you do the same passes over and over."[6] This resonated with my experience: During my 20s and 30s, I wrote six books, some in only 30 days, each exceeding 70,000 words. Instinctively, I had sharpened my skill to enter a trance-like state where the words flowed through me. Yet, as I began to venture into the world of AI, my previously reliable techniques for productivity and problem solving were seeming to stumble under its immense complexity, volume, and pace.

The very technology that incessantly calls for our unwavering attention also distracts us, conjuring an ambiance more akin to the lively Times Square than the serene Rose Main Reading Room. The abilities that helped individuals to endure and survive during the Great Depression contrast sharply with those needed for success in the AI era for one essential reason: they are mentally demanding and perpetually distracting. More alarmingly, the most crucial skill of the 21st century, required to traverse the ever-evolving landscape of AI, has taken multiple hits in the past few years, and it's set to take another one. And no, it's not attention—it's focus.

Focus and attention are intertwined abilities, and both play a role in acquiring knowledge, accomplishing tasks, and achieving goals. However, two distinctions set them apart. Attention directs your mind to a particular idea or activity while disregarding unrelated, intrusive thoughts. This encompasses being conscious of your thinking patterns, actions, and environment. Attention can also transition from one subject to another.

Focus, meanwhile, is an enhanced version of attention that involves concentrating on a specific thought or action. Times Square, social media, and AI command your attention, while the Rose Main Reading Room and adaptation require your focus. To uncover these hidden opportunities, we must explore multiple methods and take various routes to cut through the noise and zero in on the frequency of focus. Only then can we avoid the AI fatigue that threatens to render us helpless. But we've got to tip the scales in our favor because, at the moment, the odds are stacked against us.

Adults spend roughly 2.5 hours per day using social media, inadvertently weakening their concentration skills.[7] Social media not only manipulates our dopamine pathways and fosters addiction, but also significantly affects our decision-making abilities—a core skill required for success in the years ahead. A study by Michigan State University found a connection between excessive social media usage and impaired decision making when it came to taking risks, a characteristic often seen in individuals with gambling addiction and drug dependency.[8] Other research indicates that internet usage encourages attention-switching and multitasking rather than sustained focus.[9] Studies have shown that only 2.5 percent of people can manage multiple tasks simultaneously. When our brain continually shifts focus between tasks, particularly complicated ones requiring full attention, we become less efficient and more prone to errors.[10] This considerable effort leads to mental and physical fatigue, reduced task performance, and decreased concentration, known as the switch cost effect.[11] Psychologist David Meyer suggests switching tasks can cost us up to 40 percent of our productive time.[12] Other psychologists create analogies between task-switching and coordinating a dance or controlling air traffic, emphasizing that cognitive overload in these and other activities can result in disastrous outcomes.

Social media's allure, deliberately crafted to seize our attention, pales when compared with the cutting-edge integration of AI and personalization. The attention economy is evolving rapidly as commerce fervently embraces these advances to boost revenue and engagement. Traditional marketing campaigns, which once relied on broad targeting (such as urban women aged 25-30), are now being eclipsed by innovative AI-powered tools like avatars, chatbots, data analysis, and tailored content and messaging designed to captivate our minds and predict our behavior—because it has been trained on our behavior. The environment has changed, yet our mindset toward how to navigate it remains stagnant.

Unlike in the Great Depression, we are confronted with the unique challenge of dealing with the addictive allure of social media and the relentless flood of AI-related news, which together results in information overload. With countless choices and opportunities at our fingertips, we need help determining our next move, and we often delay decisions for fear they might not pan out. Our predecessors didn't experience the compulsive urge to check their social media feeds, post new TikTok videos or tweets, or constantly refresh their email inboxes. They didn't consume vast amounts of information only to turn around and repeat it across numerous platforms or find themselves trapped in never-ending doomscrolls or conspiracy theories that affected their behavior. While they faced their own terrible struggles, they possessed the mental bandwidth to adapt. The question is, do we?

In the AI era, we must clear our mental bandwidth and protect our most valuable asset—focus—so we can make well-informed decisions instead of shallow, fear-driven ones. To accomplish this, we need to look more closely at the concept of "flow" and understand how we can regain control of our focus and learn how to stay informed without being overwhelmed with information.

Mihaly Csikszentmihalyi, a Hungarian-American pioneer in positive psychology, was known as the "father of flow." His work focused on the scientific exploration of what makes life fulfilling. Though Csikszentmihalyi was not the first to recognize the flow state, he documented it as part of a broader psychological investigation. He coined the term "flow" from interviews with people who described the sensation as being carried along effortlessly by a river. Throughout Csikszentmihalyi's remarkable career, he spoke with numerous athletes, musicians, and artists to pinpoint the moments when they performed at their best and to explore the emotions they felt during those extraordinary moments of clarity. He aimed to identify the factors that stimulate creativity, particularly in professional settings, and how they contribute to productivity and problem solving. He concluded that flow is crucial for an efficient worker and indispensable for a satisfied one. Mihaly's interest in studying happiness emerged from the challenges he encountered in his early life. He experienced imprisonment during World War II and observed the anguish and distress of those around him. This sparked his curiosity about the nature of happiness and fulfillment despite adversity.[13]

In the library, I experienced what Csikszentmihalyi vividly portrayed in his extensive research. My emotions were harmonious, and I was fully absorbed in my work. I felt a sense of personal control and agency over my task; time seemed to warp and slow down, and I was able to tap into my creativity. In achieving my flow state, I had three vital components that Csikszentmihalyi had identified in his research:

1. A well-defined goal: I had invested over $15,000 to travel to Manhattan and attend the summit, aiming to secure significant media coverage to propel my career forward. I was determined to make it happen and dedicated all of my energy and brain power to this singular goal. I had spent weeks working with a former Oprah Winfrey TV producer, crafting my media pitches, and grooming myself to make the most of the 60 seconds I would have with each media representative, roughly 100 over three days.
2. I found purpose in my work, helping to teach entrepreneurs and individuals how to reach their potential. Contributing to someone's growth filled me with immense satisfaction because I had to lead by example for them to grow. That required putting myself into unfamiliar environments where I would be put to the test.
3. I was ready to push my limits. With my life's savings at risk and a feeling of stagnation setting in back home in Australia, I realized it was time for a change. Pushing my abilities to the limit and evolving my skillset became essential. I was entering the cutthroat arena of U.S media, a challenging battleground that demanded all I was ready to give.

Having achieved my goals in Australia, I eagerly embraced a new challenge in a city renowned for its grit and determination—a far cry from my childhood days growing up on the farm. But as I dealt with the stresses of the pandemic and the influx of AI, my ability to tap into this flow state became sporadic rather than routine. I found myself desperately trying to return to the Rose Reading Room. It was then that I realized that I had missed something.

Gathering my notes and laptop, I descend the library's majestic staircases until I reach the entrance, where I stand beside Patience, one of the iconic lions guarding the library's front. Enthusiastic tourists were eagerly queuing to capture memories with him and his counterpart, Fortitude. During the Great Depression, Mayor La Guardia gave them their names to represent the virtues he believed New Yorkers required to endure the crisis.[14] The names have endured: Patience watches over the library's southern steps, while Fortitude remains steadfast on the northern side.

The attention economy doesn't quite know what to make of Patience. Even though he is a hallmark of the enlightened, he doesn't seem assertive or driven enough to survive in the modern world. Yet it is impossible to access a "flow state" without his presence. Impatience, however, rejects the present, considering it flawed. Striving for a perfect future, it resists acknowledging and dealing with the now, paradoxically stunting the very change it seeks. This denial of reality leads to repeated frustration and suffering.

Contrary to common perception, patience is less about idly waiting and more about identifying alternatives and seeking opportunities. It challenges impulsive and addictive behavior that keeps us plugged in 24/7. It allows us to disconnect from the world, laying the groundwork for flow to occur and for our brains to make connections and find solutions that have previously evaded us. Maybe that is why Patience guards the New York Public Library and the Rose Main Reading Room, but more importantly, the "flow state" those who visit get to experience if only they allow themselves to.

The fast-paced tech industry leaves no room for patience, and, in our social media–driven world, neither do we. The relentless hustle culture urges us to aim higher, work quicker, and earn more money, often at the cost of our mental well-being. Social media and the rapid pace of AI rewire our brains to crave instant gratification. If we don't see immediate results, we either give up and label it a failure or, worse, place the blame on others, fostering a sense of helplessness within ourselves. But, as history reveals itself, we begin to see the connections between our past (the Industrial Revolution and the Great Depression, and more recently the pandemic), our present (social media), and our future (artificial intelligence).

In a world that often challenges patience, focus, and flow we are reminded through job loss, grief, change, or death that we need to pause, put down our phones, and stop what we're doing to reassess. By letting our thoughts settle, we can see what floats to the surface so we can choose our next steps wisely and act at the right moment. Patience also exemplifies wisdom. It shows that we acknowledge and embrace change and certain developments that need time to unfold at their own pace. But for now, we must bid farewell to Patience and Fortitude and be on our way—but our paths will cross again soon.

The second the sun hits my face from between the skyscrapers, I am abruptly reminded that midday in Manhattan is midnight back home. My brain is still on Australian time and wasn't having any of it. My brief reprieve is over, reminding me just how fragile a state of flow is. "The Roosevelt Hotel will just have to wait," I say to myself. "I need to get my luggage from storage and find my Airbnb for the night." With the summit fast approaching the next day, I know I must be in top form.

CHAPTER 3
ADVENTURER'S HANDBOOK

Rule #3: Fuel Your Focus

Reflection is a fundamental part of evolution and a key aspect of practicing patience. We can unveil critical insights into our resilience during AI-powered transformations as we evaluate our responses to previous challenges, like the sudden changes necessitated by the pandemic. The quality of our focus can determine how we use this revolution to add value to our lives. Let's explore this deeper with the following questions:

Maintaining Focus During Turbulence: A Litmus Test for AI Adaptability

1. Reflect on your response to the pandemic. How might this act as a baseline indicator for handling AI-related changes?
2. Were there aspects you wish you had handled differently?
3. Did you find it challenging to concentrate during the pandemic, perhaps experiencing an "attention recession"?
4. How often do you find yourself completely immersed in an activity and experiencing a state of "flow"?
5. Would you characterize your social media usage as excessive or moderate?
6. As you contemplate the far-reaching implications of the AI era, where do your thoughts lead you?

7. Are you an impulsive actor, or do you prefer to meticulously assess the situation before making changes?
8. If AI doesn't live up to its grand promises, what are the potential adverse outcomes of having made significant life or career changes?
9. How heavily does your business or job depend on technology, and do you need to diversify your skills or services to remain competitive?
10. What are the top three goals on your professional and personal agenda?
11. Do you have a defined purpose in life? (Don't worry if you're uncertain. We'll delve deeper into this later.)
12. Are you ready to push beyond your comfort zone, intellectually, emotionally, and creatively stretching yourself?
13. Do you have the mental bandwidth required to adapt to these changes?
14. How frequently do you engage in the practice of patience?

Decoding the Top 10 Threats of Artificial Intelligence

We now acknowledge an additional significant threat posed by AI: distraction. Much like being in Times Square, captivated by the vibrant lights and incessant noise, we can lose focus when preparing for imminent changes. Instead, we should strive to find calm, akin to the tranquility of the Rose Main Reading Room, which enables clear thought and focused preparation.

7. **Attention recession:** The captivating nature of AI, with its propensity to dominate headlines and conversations, jeopardizes our focus. This distraction can spur anxiety, mental fatigue, and the daunting prospect of further change, distracting us from what truly counts: the mental tranquility to explore possibilities with our own rhythm and time frame.

RULE #4

Open a Window

"IS THIS IT?" I wonder as I step into an apartment that is a relic of the late '80s. There is a makeshift wall separating the kitchen from the living area to create an additional bedroom—a clever adaptation to Manhattan's ever-increasing rental costs. "This is where you'll be staying," my Airbnb host Jason, informs me.

"Wow, what an incredible view," I reply. The towering, floor-to-ceiling windows make me feel like I am floating among the high-rises, with the blinding lights of Times Square just a stone's throw away.

Jason explains that this is usually his bedroom, but his work trip was canceled, so he will be crashing on the futon instead. Oh, and his housemate will be back later—apparently, he is a great guy. At this point, I just want a good night's sleep, so I decide to go with the flow. With the city packed to the brim with conferences, my options are limited: It is either this or a night under the stars in Central Park.

But as of 3 a.m., I still haven't gotten any sleep. My brain is refusing to adjust to the new time zone, and without curtains, the glare of Times Square pours in through the windows as if someone is pointing a spotlight at my face. In my earlier daze, I hadn't even noticed the bed was only covered by a single white sheet, with no duvet in sight. "Screw you, Jason," I whisper. Desperate for warmth, I rummaged through my luggage and put on every piece of clothing I could find. "Can't these Wall Street bankers afford heating?" I wonder. For some inexplicable reason, I had kept the complimentary napkin-sized blanket from my flight, which now served as an extra wafer-thin layer of warmth. As I lie here shivering, my thoughts race ahead to the next day: Would I be sharp, presentable, and able to seize opportunities, or would I crash and burn looking like a dog's breakfast on my first day at the summit? I am up against 100 others competing for the same attention, and I can't help but wonder if I could keep up with the pace. I am getting so worked up in my sleep-deprived and jet-lagged state that I imagine *smashing the window*!

If an attention recession occurs when you check out and stop paying attention, imagine what happens when you're kept up all night with questions swirling around your mind as to what your future with AI looks like. In this situation, you'll find yourself confronting the insidious eighth threat of AI. This one strikes at the core of your ability to adjust to a rapidly changing environment, both personally and professionally. This wolf scratches at the door at night, keeping you awake and preventing you from finding your flow state, stealing your attention and peace of mind. At this point, focus becomes even more vital as the productivity problem enters the equation.

Labor productivity is crucial for determining the living standards for a community, a country, and even the world. At its most basic, productivity can be defined as the amount of output produced per hour of work. As productivity increases, employee remuneration typically follows suit. During the Industrial Revolution, productivity increased due to technological innovations, resulting in job and wage growth. However, in the AI era, although productivity is expected to grow, the demand for additional workers may decline if the pace of job loss exceeds job creation, and one low-skilled worker armed with AI as their copilot could replace 10 or more high-skilled workers without it; this could push down wage growth and lead to greater economic inequality.

In a paper released by researchers from OpenAI, Open Research, and the University of Pennsylvania, their findings revealed that approximately 80 percent of the U.S. workforce could be affected by AI in some form.[1] Given the extensive reach of AI across various professions and its potential to impact a diverse group of employees, we must not overlook its effect on our mental well-being. As AI exerts unseen but real pressure on individuals to perform more efficiently, performance expectations in the workforce will increase. But will we cope?

A yearlong study by Stanford and MIT demonstrated that using AI tools, such as chatbots, resulted in a 14 percent increase in employee productivity at a technology company. This research, regarded as the first significant real-world implementation of generative AI in the workplace, involved more than 5,000 customer support agents working for a Fortune 500 software company in the Philippines. The application of AI tools for generating conversational scripts led to enhanced productivity, especially for inexperienced and low-skilled workers, who saw a 35 percent boost in work efficiency.[2] At first glance, this appears positive, but in the long run, it may also enable businesses to outsource some tasks to more cost-effective employees overseas.

Cognitive automation differs significantly from previous automation waves, and understanding this distinction is crucial. The pressure to perform and become an invaluable asset to a company or your own business as competition intensifies adds stress to an already stressed and sleep-deprived labor market.

On the other hand, longitudinal research over a decade by Cranston and Keller demonstrated that individuals experiencing flow states exhibited a *500 percent increase* in productivity.[3] When AI assistance is combined with flow, it heightens your ability to outperform others who are slow to adopt the technology and have a low tolerance for stress and adaptability. In the early months of the pandemic, one essential trait for leaders who needed to think strategically amid uncertainty was focus. Those who couldn't teetered on the edge, while many businesses, through no fault of their own, fell off the cliff. This is where technology, psychology, and biology cautiously converge as we dive deeper into our behavioral and cognitive response to change at the prospect of unstable labor markets.

To uncover answers, we must journey back to Lakeland, Florida, in 1981 to meet a young girl named Trish. Here, we will explore an issue that countless individuals will inevitably encounter yet often struggle to handle. At first, you may doubt the importance of her story, but stay with me, for I have already left you a clue.

"I wanted to kill her!" Trish's mom shockingly admitted to the judge when probed. Trish, a 15-year-old girl with dark brown hair, had her life nearly taken away by her own mother. Nine months earlier, her dad had suffered a stroke and was hospitalized. At home, Trish desperately tried to find out if he was okay. Still, her mom callously ignored her calls and pleas for information, even going as far as instructing the hospital not to disclose any details of her dad's status to the children.

When her mom finally returned home, the tension escalated as she entered through the kitchen door. "Where have you been?" Trish asked, desperation in her voice. Her mom had been shopping, bringing home carpet samples to redecorate the house. "I hate you!" Trish screamed as her body trembled with rage in response to her mom's absence of concern for her father. Then, her mom noticed a hairbrush left out on the kitchen counter. "Get this fucking mess cleaned up!" her mom howled.

To the town and her local parish, Trish's mom was adored by all. "Her skills were incredible," Trish confided. "She could make any man do anything for her." However, behind closed doors, "she was terrifying." Particularly on this day, a fierce struggle erupted, with her mom forcefully pushing Trish into the bathroom adjacent to the kitchen. With one fist jammed deep into Trish's throat and the other pressing up hard against it, her airway was crushed, her lungs gasping for air. In a stroke of divine providence, Trish's older sister Tanya—who had moved out years ago to escape her mother's abuse—burst through the door and yanked their mom off her. Tanya drove Trish to her older sister Renae's house and made the life-changing decision to move in with Renae that day. Renae, 12 years Trish's senior and blind, helped Trish prepare for court.

Nine months elapsed before Trish's mom and dad reported her missing. During this time, she lived with Renae. When the judge inquired why they hadn't reported her missing sooner, her mom replied, "We had cruises scheduled, and we didn't want to change our plans." Consequently, the judge declared Trish an emancipated minor.

As with many abuse victims, Trish tried to return to her parents' home after living with her sister for a year. But just two weeks later, they committed her to a mental hospital, where she stayed for two weeks. Upon her release, Trish chose not to return to her sister's place. To this day, she still doesn't know why. Instead, she found herself homeless, living out of her car, attending school, maintaining good grades, and working at a balloon decorating store. She would wash up in McDonalds's restrooms and occasionally sneaked into friends' homes at night, staying over without their parent's' knowledge and leaving before morning to avoid discovery.

Abuse had been a part of both girls' lives from a very young age. Trish could recall her mom forcing Renae to take scalding hot showers after beating her bloody, exposing the wounds to the searing water. Her sister's screams haunt Trish to this day. Living in a constant state of high alert, she became adept at recognizing her parents' mood swings, changes in breathing, and subtle facial movements as warning signs of impending violence.

It wasn't until Trish reached her early 50s that her husband noticed an unusual pattern: Her anxiety would escalate as the sun began to set. "I would panic," Trish said. "I'd become hyperaware of every detail in my surroundings, but I never made the connection to my past. I would start to think about where I was going to sleep for the night." Her brain had become so finely tuned to detect impending danger that a symphony of stress hormones, such as cortisol and adrenaline, would course through her body, accelerating her heart rate and breathing. Cortisol would strategically suppress nonessential systems, redirecting resources to vital areas and activating an ancient survival mechanism designed to enhance her chances of fleeing or confronting the threat. Even when the threat no longer remained, the trigger and her response did.

Trish's experience echoed what researchers have discovered: When humans find themselves in a threatening environment, such as a war zone, they switch into a hypervigilant state of mind in order to cope. However, when life treats us well, the challenges become manageable. Our ability to cope hinges on our tolerance to stress, particularly when it transforms into a relentless, unyielding force that could disrupt the economy.

Picture yourself immersed in your daily routine when you read a story on your phone that AI could overtake your entire industry, throwing thousands of people out of work. In that instant, your mind ceases to ponder trivial matters like dinner plans or gym sessions. Instead, it narrows its focus to a single element: danger. Your brain gears up to fight or flee, whether the threat is real or imagined. You find yourself engrossed in every piece of industry news that might affect your future; your vigilance is heightened to an unprecedented level.

Now imagine that this alarming news starts to circulate weekly. A colleague or a friend loses their job to AI, and if you're unprepared, you might become trapped in a state of hypervigilance. Your eyes constantly search for the lurking wolf around every corner. Your attention becomes fixated on potential threats rather than opportunities, causing you to lose sight of the present moment, the lessons you should be learning, and the work awaiting completion. How can you harness the full potential of your mind to confront the gravity of your circumstances when it is wired to focus solely on detecting peril? That reality might be closer than we can fathom.

In May 2023, outplacement firm Challenger, Gray & Christmas published a groundbreaking report that provided a detailed account of the job cuts for that month. It disclosed that AI accounted for 4.9 percent of job cuts, leading to 3,900 people losing their jobs—all within just seven months of ChatGPT's release. The report marked the first instance of AI being factored into monthly layoff data. However, the firm's representative noted that some companies might be reluctant to identify AI as the main reason behind their cuts.[4] This reluctance might result from the growing backlash against it. Insufficient reporting could create a false sense of security in employees, providing less time to prepare and adaptive measures being introduced too late.

Interestingly, during the pandemic between February and early May 2020, UK author and journalist, Johann Hari of the book, *Stolen Focus* reported a staggering 300 percent increase in Google searches for "how to get your brain to focus." [5] This increase wasn't solely due to pandemic fatigue or an attention recession; instead, it resulted from a heightened state of arousal triggered by the global alarm bells ringing at the pandemic's onset. At the beginning of the crisis, The Center on Budget and Policy Priorities estimated tens of millions of people in the U.S. alone lost their jobs.[6] Now there are indications of another unemployment crisis on the horizon as AI continues to make headlines, causing people to become hypervigilant once again. The Cleveland Clinic refers to this phenomenon as technophobia, a term describing an intense fear of technology. This fear is typically more prevalent among older generations, but AI could expand its reach, and media depictions of technology malfunctions might play a role in exacerbating this fear.[7]

Research has connected adverse mental health outcomes to increased media consumption during distressing and catastrophic incidents like natural disasters.[8] For example, the influence of media coverage on anticipated posttraumatic stress (PTS) symptoms and psychological reactions has been examined both before and after the arrival of an impending hurricane. In one 2019 study involving 1,478 Florida residents, researchers explored the relationship between forecasted PTS responses, media coverage, and mental health outcomes surrounding Hurricane Irma. The results showed a strong link between watching the news, predicted stress levels, and how people adapted after the hurricane.[9]

I still vividly remember Hurricane Irma when I was residing in downtown St. Petersburg. Our TV screens were filled with relentless news updates for a whole week, and our phones buzzed with alerts, keeping us constantly anxious due to the ever-changing forecasts. One moment we were right in the path of the Category 5 hurricane, and the next we were seemingly safe, only to find ourselves right back in danger again. This emotional roller coaster caused unusual reactions in us and many of our neighbors. Like countless other Floridians, we chose to hunker down and hold what I learned were called "Hurricane Parties," a way to distract ourselves from the looming uncertainty. The friendly competition kicked off: Who could hold out the longest without devouring their hurricane snacks before the power went out? I suspect many of us didn't triumph as champions of snack resistance.

When the storm finally passed, we emerged from it safe and without any damage to the surrounding area. As the weeklong flood of adrenaline and cortisol dissipated, it was evident that everyone was emotionally drained. I was unproductive and unfocused for the next week. Trying to catch some shut-eye when your stress hormones have skyrocketed due to the looming possibility of an evacuation order feels like an almost impossible challenge.

Living in this state for an extended period isn't just unsustainable—it causes structural changes to the brain and weakens connectivity in the prefrontal cortex.[10] This region of the brain plays a crucial role in executive function, emotional regulation, impulse control, and the ability to avoid distractions. It is also important for attention, cognition, reward, and motivational pathways. Studies have shown that during stressful times, this region essentially goes "offline."[11] And yet a large portion of society lives this way because of the "hustle-until-you-die" culture.

This phenomenon isn't exclusive to hurricanes, work, or other traumatic events, as even watching violent news stories on social media can cause individuals to experience symptoms similar to those associated with post-traumatic stress disorder (PTSD).[12] Regardless of whether it's PTSD, hypervigilance robs us of our capacity to concentrate on the broader perspective, leaving us frozen by the whirlwind of information that surrounds us. This state of heightened alertness is a striking manifestation of hyperarousal, where your attention becomes intense and nearly impossible to disengage. It is exhausting and detrimental to our strategic thinking skills, further emphasizing the importance of understanding the impact of AI and the news media on our psychological well-being.

THE WOLF IS AT THE DOOR

This adrenaline-fueled reaction, commonly referred to as the fight or flight response, originates from the activation of our sympathetic nervous system. It triggers a stress reaction within our bodies and narrows our focus, preparing us to either face a threat or escape from it. A narrowed focus can be incredibly effective in a safe environment and life-saving in a dangerous one. But when your fight-or-flight response remains stuck in the "on" position, it can hinder your ability to engage in sustained forms of concentration, which is essential for effectively planning and navigating your future, especially if that future appears uncertain.[13] Incessant news about job losses can push some people to smash their "window of tolerance" and enter a state of hypervigilance, similar to the one Trish used to survive her abusive parents.

This window of tolerance, introduced by Dan Siegel, a clinical professor of psychiatry at the UCLA School of Medicine, is essential for understanding how people react to stress and change. It refers to the optimal range in which a person reacts to stimuli without excessive arousal. This range helps us naturally regain calmness and maintain emotional balance, representing a psychological and biological state that enables us to "decompress" after experiencing stress.

Our prefrontal cortex and executive skills are at our disposal when we are within our window, allowing us to plan, prioritize complex tasks, maintain focus, regulate emotions, manage time, and, most importantly, solve problems. This empowers us with the necessary tools we need to adapt. Two other zones flank the "optimal arousal zone": hyperarousal and hypoarousal. Optimal arousal embodies a state of emotional adaptability, curiosity, presence, and the capacity to manage life's stressors effectively. However, when stress triggers us, we may smash the window and enter either hyperarousal, an emotional state marked by anger, panic, irritability, anxiety, and hypervigilance, or hypoarousal, which is defined by emotional shutdown, numbness, withdrawal, shame, burnout, and disconnection. We lose access to our prefrontal cortex and cognitive resources in either state.

OPEN A WINDOW

Like the flow state, hyperarousal distorts time; however, instead of slowing it down, it accelerates it, triggering anxiety. In contrast, hypoarousal causes time to move at a painfully slow pace. Ultimately, remaining within the optimal arousal zone is essential for leading a functional, healthy, and happy life. Nonetheless, life doesn't always make staying in this zone easy, and it's natural for us to fluctuate in and out of it. The objective is not to remain within it constantly, as that would be unrealistic. Instead, we should strive to broaden our window of tolerance and enhance our ability to demonstrate resilience by swiftly returning to our optimal arousal zone whenever we find ourselves adrift outside of it.

Fascinatingly, the window of tolerance closely aligns with Csikszentmihalyi's flow state. He wrote, "Enjoyment appears at the boundary between boredom and anxiety when the challenges are just balanced with the person's capacity to act."[14] If anxiety represents hyperarousal in Siegel's model, then boredom, as Csikszentmihalyi refers to it, in more extreme cases, represents hypoarousal, commonly referred to as burnout. It's a collapse of our ability to cope.

At a time when we are particularly vulnerable, AI's emergence has intensified an existing societal issue: a work culture that pushes us to work harder, smarter, and faster, often at the cost of sleep, health, and relationships. This results in a society that is increasingly hypervigilant and prone to burnout. The pandemic exposed our personal windows of tolerance in various ways, frequently resulting in shorter tempers and reduced patience for change. It's essential to understand that everyone has a unique window, influenced by a myriad of factors such as past traumas, physiological makeup, and daily stressors. Like fingerprints, no two windows are identical. However, your past doesn't always have to determine your future, even when facing extreme odds—especially if you can learn to become finely attuned to your behavior.

"When my mother died, oh my God, I won!" Trish said emphatically. "I didn't let her anger, bitterness, or my terrible upbringing stop me from loving people, no matter what she put me through. I still managed to care about others." Trish's mom was 93 years old when she passed away, and a common joke among her family was, "Evil doesn't die." What was her secret to getting through this tumultuous time? Only when she was ready, she unpacked her past one piece at a time. Along the way, she collected tools that allowed her to plan her future while revealing her strengths and weaknesses. Only then could she be free to move forward. But countless souls remain shackled to the memories of their past, unable to envision a different future that lies just beyond the noise and chaos. They are held captive by the echoes of yesteryears, blind to the possibilities that await them in the uncharted realms of tomorrow. This inability to break free from yesterday ultimately hinders their capacity to adapt and embrace tomorrow.

Regardless of your experiences with generational trauma, everyone must dedicate time and effort to remain within their window of tolerance, allowing them to achieve a flow state amid technological advances that threaten to disrupt it. Our news consumption has grown disjointed, resulting in a fragmented worldview pieced together through shreds of information, hindering our ability to detach from vigilance or move past avoidance. Hypervigilance and hypoarousal arise when we cannot solve a problem and fall back on coping with it. Intriguingly, exposure presents a dual role as both a challenge and a potential solution to the eighth threat of AI: hypervigilance.

"Good heavens, it's 7 a.m. already!" I exclaim, caught off-guard. At this point, the bright lights of Times Square shining through the window have been replaced by the gently rising sun. I have only 30 minutes to spare before I must leave for the summit. My restless thoughts, the glaring lights, and heightened vigilance in an unfamiliar setting had conspired to keep me awake for most of the night. I leap into the shower, hastily dress, and bolt out the door with my trusty brown leather satchel in hand. Racing toward Starbucks for the third time, I buy a steaming black coffee and gulp it down at breakneck speed, praying it will jolt me out of my current blend of exhaustion and frenzy. This is a far cry from the media pitches and interviews I'd done back home in Australia; I was now among the heavyweights of the media world. As I enter The Roosevelt Hotel and register, to my relief, I encounter a familiar face from back home: marketing consultant Cydney O'Sullivan. Although we have only ever communicated via Skype, our connection was instantaneous. Both of us were a bundle of nerves, even though we'd spent months preparing for this event.

I have never been in this type of high-pressure environment before, but here I was, fully committed. I've journeyed more than 10,000 miles to be in this place. The media producers don't care whether I am tired or not; their only interest is whether I have something valuable to offer them. Exposure is what I want, and that is what I will get. I just may not get it in the way I expect.

CHAPTER 4

ADVENTURER'S HANDBOOK

Rule #4: Open a Window

As we further investigate the trials and possible solutions within our dynamically evolving workspace, we must contemplate a set of discerning questions designed to reveal vital concerns relating to mental health, workload pressures, and the assimilation of automation and AI into your professional life. By clearly articulating your apprehensions, you can spark innovative thinking and resolve, enabling a thorough evaluation of the necessary dialogues, time commitments, and resources required for effective problem solving.

Growth Amid Disruption: Cultivating Tolerance for Change

1. What anxieties or issues hinder your ability to enjoy a peaceful night's sleep?

2. What immediate actions can you implement to mitigate those concerns?
3. Does the amount of your work cause you a state of extreme stress (hyperarousal) or diminished responsiveness (hypoarousal)?
4. Which day-to-day habits help keep you balanced and within your window of tolerance?
5. Can you identify cognitive tasks in your job that could be automated to alleviate workload strain without putting your job or business at risk?
6. How do you envision AI enhancing your job performance when it becomes part of your work environment?

Decoding the Top 10 Threats of Artificial Intelligence

We are now becoming aware of the eighth threat posed by AI: the triggering of hypervigilance. Comparable to echoes from past experiences and traumas that resurface when least expected, the incessant and fast-paced evolution of AI has the potential to stir up these latent patterns, complicating our adaptation to this transformative era.

8. **Hypervigilance**: AI's rapid progress might induce a state of continuous alertness, placing us in a reactive state rather than a proactive one. Existing in a world where change is the only constant, we may find ourselves reverting to old behavior patterns in response to these abrupt shifts, destabilizing our mental equilibrium and hindering our decision making. Consequently, we might find ourselves entangled in past reactions and future apprehensions rather than fully leveraging the opportunities offered by AI.

RULE #5

Accelerate Adaptability

I SOMETIMES FIND MYSELF MARVELING at my dad's remarkable resilience and adaptability. Life threw him some devastating curve balls in his younger years. First, his brother Bruce lost a fierce battle with leukemia at the tender age of 16. Tragedy struck again barely two years later, when their father lost his battle with a brain lesion at 42. These heart-wrenching events sent my dad into a whirlwind of despair, facing increased responsibilities on the farm while turning to alcohol to numb the pain. One unforgettable incident saw him being stopped by a police officer for riding a horse while drunk. As the officer attempted to issue a citation, my dad defiantly retorted, "It's a horse, not a car," as he galloped away. Yet, there was a pivotal incident that not only brought his reckless behavior to an end, but also resulted in a broken back.

One night, on a desolate outback road devoid of cars, a deadly mix of alcohol and fatigue overcame him, plunging him into a heavy sleep as he drove. Suddenly, his car careened through a fence and somersaulted into a waterless dam. Dawn broke, casting light on the wreckage, when they finally discovered him. Miraculously, he defied the odds and made a complete recovery.

His experiences with grief, loss, and suffering made him a resilient man. He became intimately familiar with the whirlwind of change, even when it exacted a steep price. As a merciless seven-year drought descended upon our farm in the mid-1990s, my father sprang into action, diversifying our income streams and conversing with fellow farmers to discover emerging trends. Unbeknownst to him, Steele embodied the essence of Stoicism in the face of adversity.

In the bustling ancient Greek city of Athens, around 300 BCE, Zeno of Citium laid the foundation for the enduring philosophy known

OPEN A WINDOW

as Stoicism, an ethical philosophy and practical approach to seeking wisdom in life, which revolves around the idea that our reactions stem not from actual events, but from our interpretations of them. Ancient Stoic thinkers emphasized the importance of distinguishing between things in our lives that we can control and things we can't, urging us to relinquish our worries about the latter, even if that task is arduous at times to do.[1] Through this approach, we are better positioned to navigate the unpredictable tides of existence.

To moderate our responses to the often startling ups and downs of life, we must deliberately expose ourselves to experiences that enhance our resilience, promote familiarity, and nurture adaptability, both to improve our stress response and expand our window of tolerance for change. Adaptability is an essential skill forged in the fire of life's experiences. It entails processing new information and circumstances and then modifying your actions accordingly. Both humans and AI demonstrate this exceptional capacity for growth, often achieved via hard-earned lessons and persistence (and coding, in the AI's case).

An adaptive AI model can persistently learn and evolve as it is used, implying that over time, identical queries may yield different outcomes as the model refines its problem-solving approach. A fixed model, on the other hand, undergoes training, development, and testing to achieve its optimal version. Once this model is deployed for public or private consumption, it consistently generates the same results whenever the same input is provided, much like a calculator.[2] Just as adaptive AI continually evolves and learns, so do individuals with a growth mindset, thriving when confronted with challenges and embracing change. Conversely, those who resist transformation and demonstrate a lack of resilience are akin to fixed AI models, locked into a rigid mindset that hinders progress.

Embracing a growth mindset empowers you to venture into uncharted territories. This mindset cultivates resilience, allowing you to rebound from inevitable setbacks and obstacles. It encourages you to refine and enhance your personal "product"—yourself—adding unique value as an employee or entrepreneur that AI will have difficulty replicating. And it nurtures humility, as you remain open to continuous learning and growth, steering clear of stagnation.

In a similar vein, both humans and AI depend on the input they receive and their exposure to patterns and behaviors from various sources, especially their environment. By seamlessly integrating these factors, they demonstrate an exceptional capacity for adaptation and evolution. Yet as the West finds itself grappling with the challenges that will hinder our ability to adapt—including information overload, attention scarcity, hypervigilance, and AI integration—we see China forging ahead with a well-defined vision of the future. China has deployed three strategies thus far, along with one more that has been tried and discarded. By examining and integrating these methods into our own approach, we can forge a strategy that enhances our adaptability at a significantly faster rate. The first of China's approaches was the use of neurofeedback combined with meditation.

As early as October 2019, *The Wall Street Journal* reported China's unwavering commitment to preparing its children by investing billions into AI.[3] In an innovative first approach to education, a primary school in China introduced headbands that gauged the focus of each student in the classroom. This groundbreaking technology instantly sent the data to the teacher's computer and parents, providing real-time insight into the young learners' engagement levels. Red and green lights on the headband alerted the teacher to when the student was focused or distracted. Armed with this information, teachers could tailor their approach to better align with the unique needs of each student.

In the pilot, students would begin their day by donning these devices and engaging in mindfulness exercises through meditation, thus combining the wisdom of the past with 21st-century technology.

However, three months into the program, public outcry over potential infringements on the children's privacy caused its abrupt termination. The education bureau of Jinhua City explained that the background data was solely intended to aid teachers in analyzing and enhancing their teaching methods, and they only used it once or twice a week for up to 30 minutes.[4] However, as of 2023 the program has not been restarted.

It remains unclear whether there was misuse or only a perceived misuse of the technology in this case. As we continue to explore this subject, it will become evident that neurofeedback may deliver a significant advantage when appropriately used. For consenting adults, such as us, it could provide an edge that Chinese students might have missed out on amid the controversy. Yet, as with any technology, not all their methods are without limitations or concerns, setting the stage for their second approach: AI tutoring.

Innovative companies such as Squirrel AI have crafted custom tutoring systems that use AI algorithms to tailor lessons. Children in China who have used the systems have raised their test scores significantly, but some educators are concerned that the accelerated adoption of AI in education may place excessive emphasis on standardized testing and unrealistic expectations placed on the participants, leaving students ill-equipped to navigate a rapidly evolving world.[5] What is clear, however, is that China is tackling AI head-on by exposing youth to it early through their third approach, AI literacy programs.

The definition of AI literacy is constantly changing. According to Bryan Cox, head computer science specialist at the Georgia Department of Education, it encompasses exploring technical aspects, such as understanding the inner workings of AI, its functionality, and its effects on individuals and the environment.[6] Possessing AI literacy is imperative, as businesses will need teams with fundamental knowledge of AI's practical applications in the corporate and private sectors. It also plays a crucial role in understanding how, and when you may need to pivot in your career.

Their fourth approach involves implementing censorship and other restrictive measures. Take Douyin and TikTok, which are two sides of the same colossal social media app coin. Douyin, the original version released in China, paved the way for its global counterpart, the wildly successful TikTok. The two platforms showcase distinct flavors of content. Douyin thrives on educational material, with videos aimed at enhancing skills and personal growth to upskill and protect children's mental health.

Meanwhile, the global audience gets the dopamine-addictive version, TikTok.[7] According to the *South China Morning Post*, in 2021, ByteDance, the parent company of Douyin and TikTok, implemented a compulsory five-second break after a user spent a long time watching videos in China. The measure was aimed at minimizing the risk of addiction; it's unclear how long the viewer had to watch to trigger the pause.[8] The lasting impact of these technologies on the nation's 200 million children will gradually unfold as they mature into adulthood. But should we wait until we see the results before acting?

It is evident that various strategies and measures have been and are being attempted to maintain competitiveness—to recap, neurofeedback and meditation, personalized AI education, restricted social media use, and AI literacy. They are skillfully upgrading their AI and concurrently elevating their population's ability to adapt. In contrast, the West is upgrading its AI while downgrading its population by not providing the necessary resources for adaptation.

China's approach is not unattainable. When done at a personal level without infringing on fundamental human rights, it is a powerful arsenal. But there is a mindset that will prevent its adoption, shrinking the window of tolerance and ultimately leading to a rigid, unyielding belief system. Its name is splitting, and all of us have been guilty of it at some point in our lives.

Since 2020, the world has been swept up in a tide of binary thinking, with situations and people being labeled as either good or bad, black or white. Splitting, a psychological defense mechanism that emerges as a subconscious response to help individuals cope during turbulent times, can help unveil a deeper understanding of our responses to AI.

OPEN A WINDOW

The evolution of this phenomenon can be associated with traumatic experiences early in life, including maltreatment and desertion. Individuals grappling with splitting often tend to experience intense emotions and find it challenging to accept that positive and negative qualities can coexist simultaneously. This protective measure shields them from intense negative feelings, such as insecurity, abandonment, and isolation, providing the illusion of a safe harbor during upheaval.

Social media algorithms reinforce splitting by weaving a captivating spell of division, presenting content in users' feeds that they are most likely to click on or engage with, based on their previous interactions. Contrary to popular belief, we significantly influence the algorithms; a simple experiment of focusing your interactions on educational content for a week will reveal how the algorithm adjusts to prioritize similar content, creating a varied input that broadens your thinking and creativity. This exposure to a fresh perspective empowers you to explore social media differently. However, it requires personal effort and adaptability, testing your emotional regulation. In the initial months of my immersive exploration into AI, I couldn't shake the feeling that I was failing this test until I came to an essential realization: Despite its complex tapestry of benefits and drawbacks, AI is neither inherently good nor evil. But public opinion on this subject is splintered. Much like the virus that polarized us during the pandemic, or the vaccine that saved us from it, AI has become the new catalyst for division. The further I delve into its possibilities and threats, the closer I come to one conclusion: No one position on this spectrum is exclusively advantageous; instead, a dynamic balance between skepticism, optimism, realism, and doomsaying is crucial for adapting in the years to come and for helping to establish a balanced policy that benefits us all.

Holding stubbornly to a single position, regardless of new information, is emotionally and mentally draining. It deregulates and damages your ability to adapt, often prompting individuals to react in extreme ways and contributing to social discord and division. Once you understand why people act the way they do (i.e., lashing out as a coping mechanism because they deem change as outside of their realm of control and window of tolerance), it reduces your need to fight back, saving much needed mental bandwidth for adaptation. If we fail to understand this basic premise, we risk becoming as uncreative and rigid as a calculator and as emotionally stable as a teenager.

As the importance of seamlessly transitioning between diverse perspectives increases, adopting a growth mindset and emulating adaptive AI techniques that are inherently human becomes essential. By continuously refining our understanding of fluid problems and adjusting our approach, we allow new input to generate fresh solutions and can face surprising revelations without shattering our sense of reality. Instead, our perspective expands to encompass the evolving tapestry, empowering us to change course as needed. The importance of expanding our perspective becomes evident when we acknowledge our innate inclination to seek clear-cut answers, as they provide certainty and reduce ambiguity. AI challenges us to find solutions within our sphere of control and window of tolerance while filtering out distractions, particularly when achieving community consensus appears unattainable. But does this mean we should accept or tolerate every new development in AI without opposition?

No. By managing our emotions, we prepare ourselves for a sustainable journey, allowing us to avoid burnout. Acknowledging our sphere of influence enables us to have an impact. Public opinion on AI can create change and protect workers' interests, once you have the attention of those pursuing a balanced approach. The current situation mirrors the Industrial Revolution, when workers' demands for rights starting in the late 19th century led to lasting benefits that we still enjoy. The more individuals engaged in this conversation, the better.

The Writers Guild of America may represent one of the initial movements of resistance, demonstrating the potential of union power in combating AI exploitation when they went on strike from May to September 2023. Their central message was a rejection of the idea that

AI-generated content constitutes "the future" and a statement that writers must be protected.[9] They claimed, among other things, that studio executives were planning to use AIs to write the first drafts of scripts—cobbled together from thousands of works written by WGA members—and then hire union screenwriters to rewrite the scripts at a much lower rate than they would be paid for writing an original work.

The strike stretched on for almost five months and shut down dozens of film and TV productions as negotiations stalled, including the final season of *Stranger Things* and *House of the Dragon*, HBO's *Game of Thrones* prequel.[10] The writers seized what lay within their realm of control and transformed it into a formidable weapon—and used it to win a real victory over AI.

The final contract was the first major union contract to set out real limits on the use of AI. According to the contract, studios are allowed to use AI, but barred from using it to replace writers or cut their pay; writers cannot be compelled to use AI in their work; and, perhaps most significantly, while studios are still allowed to use writers' work to train AI, "the WGA reserves the right to assert that exploitation of writers' material to train AI is prohibited by MBA [their collective bargaining agreement] or other law."[11]

The choice ahead of us lies in the delicate balance between creative expression, economic gain, and societal impact. For countless entrepreneurs, leaders, and managers, financial pressures may force their hand, leaving little room for personal preference for humans over AI. As adaptive AI becomes as ubiquitous as email or phones, we may find ourselves having to confront feelings of guilt for actively participating in training systems that displace employment opportunities for others, especially future generations.

Despite that daunting prospect, the responsibility for shaping the future doesn't solely rest on big tech, policymakers, or us as individuals. Each of us possesses the ability to influence the outcome, and no one is exempt. We all actively contribute to building the future and adapting for the betterment of humanity. When we let our thoughts become chaotic and unfocused, we risk getting stuck in a fixed mindset rather than embracing the broader implications for humanity. This is exactly why, despite the initial failure of neurofeedback integration within China's diverse AI strategy, meditation remains a common practice among students in various schools. It's one that many will overlook as being too simplistic.

As the pandemic intensifies, my daily meditation becomes an unwavering commitment, a necessary sanctuary amid the chaos. However, it becomes increasingly harder to get back to the Rose Main Reading Room and my flow state, but I must find a way. My cortisol levels have risen, my sleep is disrupted, and my thoughts are a chaotic storm. If I can devote that much time to social media, which is consuming me, I can dedicate an equal amount of time to an activity that counteracts the damage caused by information overload and constant change. I'm glad I made that commitment, as the gut punches come hard and fast.

Just as our business started gaining momentum, fate swept us off our feet, forcing us to adapt again. Looking back, the escalating setbacks we faced seem almost comical, and it wasn't just the massive drop in income that caused me to become hypervigilant. My story is much less captivating; in fact, it's downright embarrassing: I tripped on the steps at Donut King on a weekend trip to St. Petersburg Beach, hurting my back. It's the most expensive donut I have ever purchased, costing more than $3,000 in medical expenses and almost three years of excruciating pain. Yet, every attempt I made to heal and get back on track was thwarted. Following a chiropractic appointment that aggravated the pain, a cryotherapy session caused it to drop from my upper back to my lower spine, making it feel like someone had reached inside me and was crushing it. Within a few short months, I was bedridden, unable to walk for more than 10 minutes without being in agony. I spent my 40th birthday lying flat on the floor in pain. Yet these were only the whispers of the challenges that lay ahead. We couldn't predict what was going to happen next.

OPEN A WINDOW

In renewing our U.S. visa, our petitioner had a sudden change of heart the night before, leaving us in a precarious situation where we might have to pack up our entire apartment, say goodbye to our beloved rescue dog Mitch, and leave for Australia with less than 10 days' notice. I still recall the day we received the news.

"What are we going to do!?" I ask my partner. My heart races as I teeter on the edge of a panic attack. Mitch, a 5-pound Yorkie with a tongue that perpetually hangs out of his mouth, leaps onto my lap, and scampers up my chest, eagerly licking my nose and wagging his tail with boundless enthusiasm. I had been his emotional support human since rescuing him; we huddled together in the closet as we weathered the Florida storm season. Now the tables had turned—my little rescue dog had become my savior, breaking my emotional pattern of distress and empowering me to face our new situation with determination. I embarked on an ambitious mission: signing two book deals on the same day and completing both within six months. Miraculously, I pulled it off. In an attempt to escape the increasingly crowded downtown life, we signed a lease for a three-bedroom apartment by the water, but two weeks before our move-in date, we were informed of construction delays. Our only option was to live in hotels in Orlando—during spring break.

While I was dealing with crippling pain, a body riddled with inflammation, immigration problems, a difficult business environment, writing two books, and living out of a suitcase, I needed a way to counteract the negative thoughts of doom and gloom. That meant more than meditating once per day; I had to increase it to three times a day and incorporate neurofeedback. I had to train my brain like an athlete trains their body for a marathon. Put simply, the solution must rise to the level of the challenge. But we must also fight our bias for equating complexity with effectiveness and results, as complex solutions are often hard to implement and adhere to. We must probe deeper and examine whether our networks require an upgrade, akin to China's efforts to help their students adapt.

THE WOLF IS AT THE DOOR

To grasp AI, simplify its complexities, and understand its practical applications and implications, we must accelerate our learning process. To achieve this, our brains must be in a state conducive to acquiring new knowledge. This requires maintaining a sharp focus. Among the multiple strategies China has tried in its schools, meditation and the combination of meditation with neurofeedback, regardless of the controversy there, should get the most attention from you.

Neurofeedback, often referred to as EEG (electroencephalogram) biofeedback, is a therapy that relies upon real-time evaluations of an individual's brainwave patterns via a specialized computer program. This process uses either auditory or visual cues to guide users in identifying and adjusting their mental processes. This unique approach empowers individuals to control and enhance their brain function, potentially easing symptoms related to neurological and mental health disorders. For instance, some devices can convert your brain activity into weather-related sounds. When your mind is wandering, you'll hear the sounds of a storm, but as you recover your focus, the weather becomes tranquil, accompanied by the sound of chirping birds. It reinforces the habit of focus while actively reminding you what focus feels like, allowing you to maintain concentration for extended periods, and teaches you to enter flow on command. Much like executing a single rep during a workout, you're learning to strengthen your "focus" muscle. It's the gamification and democratization of brain training—a tool fit to compete or collaborate with AI.

I first explored this technology in 2018 when researching my book *Unstoppable*. At the time, home-use devices could be purchased for around $300. After just a few days of practicing neurofeedback and meditation with one of these devices, I observed a substantial increase in my calmness. My emotional responses were more managed, situations that would usually unsettle me slid away like water off a duck's back, and my focus sharpened to a laser point. The combination of meditation and neurofeedback, in addition to improving your focus, also offers a promising solution to hypervigilance. In a 2023 study by the Mayo

Clinic, neurofeedback reduced stress and burnout among health-care professionals by 54 percent by enhancing mindfulness techniques. Additionally, an impressive 91.9 percent of participants reported feeling more relaxed after using the device. These results emphasize the enormous potential of wearable technology in helping individuals in high-stress situations protect their mental health.[12] Furthermore, studies have consistently demonstrated that neurofeedback and meditation enhance working memory capacity[13] and contribute to increased gray matter volume, which supports brain regions associated with learning, memory, cognition, and emotional regulation.[14]

Experts also suggest that meditation could alter the brain's structure, enhancing mental activity and cognitive flexibility and making your brain stronger after just eight weeks.[15] Additionally, it can decrease activity in the amygdala, which is responsible for fear, anxiety, and stress.[16] A meta-analysis even demonstrated that its effects rival antidepressants for depression and anxiety.[17] However, the most important finding of the research into meditation is that it supports neuroplasticity.

The notion of neuroplasticity has existed for more than a century, yet neuroscientists didn't fully embrace it until the mid-20th century. Neuroplasticity represents the brain's extraordinary ability to adjust and transform its structure, functions, and connections in reaction to internal and external stimuli and is intrinsically linked to our conscious and subconscious exposure to experiences.[18] Adapting and learning at a competitive pace without maintaining your neuroplasticity will become inconceivable. Interestingly, research has found that children who are blind demonstrate improved connectivity and reorganized neural pathways when compared with sighted children. This suggests that the brain makes up for the absence of vision by modifying its structure and function, enabling blind children to more effectively process information from their remaining senses, like hearing and touch.[19]

However, as we face the overwhelming influx of social media, constant change, and heightened vigilance, we are approaching the boundaries of our adaptability. To overcome these limits, we must change our trajectory and reintegrate time-honored practices into a technological and biological stack that functions in harmony. These practices include China's stack, neurofeedback, meditation, AI literacy, and carefully regulated social media consumption to counteract the adverse effects of life in the 21st century. If we don't, we limit our capacity to adapt and consolidate new information and form critical connections that unlock our future, similar to what I attempted to achieve at the summit back in Manhattan.

"Phew!" I whisper as I scan the day's schedule, feeling relief wash over me. Today we are going to absorb the wisdom of various speakers; we don't have to face the media until tomorrow. I still have a chance to fine-tune my thoughts and brain, but it will be a challenge. Never has jet lag gripped me so fiercely. Desperately fighting to keep my eyes open as noon approaches, I drift in and out of focus, determined to maintain control. I have just one shot to make this work. Back in my early 20s, I had mastered a unique technique akin to meditation that rewires your neural pathways to guide you toward your goals. This technique once helped me conquer the unimaginable. But before I can deploy it, I must peel back my sense of self—something that many of us will need to do in the years ahead. It isn't easy or comfortable, but it is necessary if you want to adapt and refocus.

CHAPTER 5
ADVENTURER'S HANDBOOK
Rule #5: Accelerate Adaptability

The rapid evolution of AI presents us with the challenge of swiftly adapting and refining our skills to stay relevant. Embracing change and exploring diverse methods of reskilling and upskilling becomes integral to our success. By adopting a fluid mindset, we can remain malleable and resilient in the face of transformations. Moreover, drawing inspiration from China's approach provides valuable insights into how to seamlessly integrate AI into our existing systems.

Dancing with Disruption: Mastering the Art of Adaptability

Here are seven vital questions to contemplate as we harness our ability to adapt and thrive in a world driven by AI:

1. How can you foster a growth mindset in your professional life?
2. Can you easily navigate multiple perspectives to gain new insights, even when they challenge your current beliefs?
3. What steps are necessary to enhance your competitiveness in your domain?
4. How do you exhibit adaptability and versatility in your skill set?
5. What crucial discussions must occur in your workplace regarding AI policies and limitations?
6. How can you incorporate the four approaches for AI adoption: 1. Neurofeedback and Meditation, 2. AI tutors, 3. AI literacy, and 4. Restricting social media access to regain focus?
7. How can you best familiarize yourself with some of the tools that are being used in your industry?

RULE #6

Embrace Reconstruction

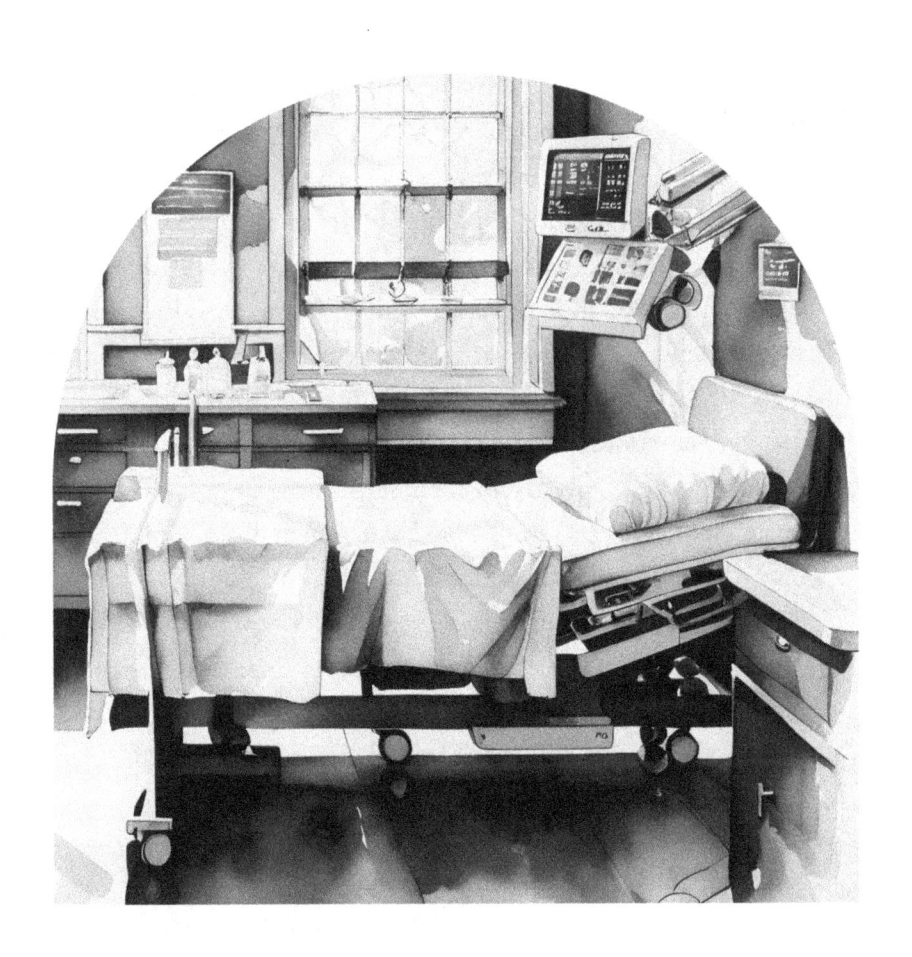

IN THE HUSH OF THE HOSPITAL ROOM, my mother sleeps peacefully on a bed just a few steps away from me. I park myself on a rigid, cold, cream-colored chair. The room, dimly illuminated by light seeping in from the hall, contrasts with the unrest in my heart. Eventually, I, too, fall into an uneasy sleep.

Abruptly, around 2 a.m., a noise from the far side of the other bed jolts me awake. Now 53, my father struggles to communicate. A full two weeks have passed since he has last spoken. My mother remains by his side night and day in the palliative unit. I rise quietly from my seat, mindful not to break her slumber, and walk toward Dad's bed, afraid that this was the inevitable moment we all feared.

As I approach, I notice Dad's lips are dry and cracked from the arid air in the air-conditioned room; he needs water. I fill a glass to the brim and drop in a blue straw, and then gently raise it to his lips. I wait a few beats before setting it down and pull a heavy chair closer to his bed. I enfold his hand in mine as we exchange silent gazes. He gently squeezes my hand back as if to say that everything will be OK.

Once he drifts back to sleep, lulled by the high doses of morphine he is on to dull his pain, I pull out a piece of paper and scribble down all the goals I still want to achieve. I keep going for the next two hours until exhaustion overtakes me. This is more than a to-do list; it symbolizes my pledge to him and myself — to use the gifts he's given to me.

That was the final moment I would share with my dad, who died shortly afterward. At 23, I resolutely set my emotions aside and became a bastion of strength for my mother. Her love story spanning 37 years of marriage, with the "love of her lifetime" was over.

EMBRACE RECONSTRUCTION

Rewind to a month earlier: My dad had had another run-in with a fence on the farm. This time, he wasn't embroiled in a struggle with his tractor or drove through one intoxicated in his car, but was tightening the fence with a ratchet strainer to prevent the cattle from escaping. The tension grew so much that the fence suddenly snapped and the heavy metal lever struck him forcefully on the head, knocking him out cold and giving him a concussion. In the subsequent weeks, Mom noticed a dramatic change in his behavior as he exhibited uncharacteristic forgetfulness. A later brain scan revealed something unexpected: He had a brain tumor about the size of a golf ball, located at the center of his brain. It had been growing for months.

We decided as a family that he would undergo surgery to see if it could be removed. Unfortunately, the surgery proved problematic. Initially, the prognosis was three months, but he didn't make it beyond four weeks due to unforeseen complications.

The day I buried my father was the day I buried my sense of identity. Bereft of his guiding presence in my life, I felt a deep sense of irrelevance. At 23, I was still discovering myself while embarking on a new career. Yet there were promises I had made to both of us on that piece of paper in the dark, quiet hours of the morning that I planned to keep. I kept that piece of paper framed above my desk, where it served as my North Star. Among the many ambitions scribbled on it, one stood out in particular: "Visit New York." I'm not sure why, but the city held a peculiar allure. However, life isn't always linear; it's an interwoven journey of experiences. Setting foot in New York would have to wait as I embarked on a somewhat different expedition, navigating my way through the seven stages of grief. Unbeknownst to me then, this phase of my life would lay the foundation for today.

Fast-forward to 2023, and I can see a striking similarity between my journey through grief and humanity's reaction to AI, epitomized by the ninth threat: irrelevance. Contemplating existential questions like, "What is my purpose in life if AI can perform my job better than

I can?" helped me decipher this interconnected puzzle. I realized that understanding our role in an AI-driven world might depend on our relationship with grief and, more important, how we reshape our personal and professional identities in its wake.

It was once described to me that the stages of grief resemble a string of visitors occasionally dropping by to sit with you. Sometimes they visit simultaneously, while other times they recede into the background or lurk behind a veil of denial. Elisabeth Kübler-Ross, an American-Swiss psychiatrist, initially introduced the concept of the five stages of grief back in the 1960s.[1] Over the years, her original proposition has been expanded to seven stages. Placing the matter within an AI context, however, takes us into unforeseen territory.

The initial stage involves shock and denial, characterized by an overwhelming sense of disbelief when confronted with change or risk to their livelihood or safety. As the prospect of becoming irrelevant looms, the struggle to comprehend their purpose in life intensifies. Yet, despite emerging trends and patterns that indicate they may next be in line to become irrelevant; they deny the possibility. Next comes pain and guilt—they feel remorse for thriving while others endure hardships or fail to adjust to changes until it is too late—or significantly harder to do. The third stage is marked by anger and bargaining: they may strike back against impending changes or resort to bargaining by lowering prices to retain clientele or working themselves to the bone to prove their worth and keep their jobs. The fourth stage ushers in depression. At this point, they withdraw from others and grapple with the prospect of losing their career or business, to which they've devoted their entire life. They mourn the loss of the familiar and question the purpose of existence. The possibility of having to start anew, especially after losing it all, becomes the source of frequent restless nights. Each phase plays a significant role in the process. Denial, for example, is an understandable reaction when faced with daunting challenges like a proverbial wolf at your door. But living in denial can only work for so long. Sooner or later, you come to a crossroads and the first of many choices that shape you—and act like rocket fuel for focus.

"Why am I aching all over?" I ask myself, shivering uncontrollably, seemingly out of nowhere. "It's not even cold in here." At 3 a.m., I crawl out of bed, unable to stand, my limbs betraying me. Inch by inch, I drag myself down the wooden hallway, and finally make it into the bathtub. I shakily turn the hot water faucet on, still quivering as I immersed myself.

Oddly, I feel relatively intact emotionally, but my body is singing a different tune. These nighttime tremors have haunted me for a week. By day, I go to work and handle my usual tasks. But I have reached my breaking point, so I march to the doctor's office in Melbourne the next day, desperate for answers. The doctor looks me in the eye and says, "Ben, you're experiencing grief. Your body goes into shock at night when your subconscious seizes a moment of stillness to process the trauma." I am stunned by the revelation, words escape me.

For three months, I had sidestepped my grief. Moreover, I wasn't just dealing with the loss of my Dad. I felt utterly alone, grappling with life and entrepreneurship, trying to rebuild everything, including myself. Doubting my worth and identity, I soldiered on, feigning normalcy. Desperately trying to cling to my past sense of self, I waged a war between my past, present, and future until I found myself trapped, incapable of envisioning my next move. And that's when it hit me.

Many of us intertwine our identities, relationships, and careers so tightly that disentangling them becomes painful. Ask yourself: Who are you if you are stripped of your job title or ability to work in your field? Take a moment to reflect on your personal growth, and consider that the person you are today might not resemble the one you need to evolve into for the future. This was starkly illustrated in a *Washington Post* article about Eric Fein, a content writer in Bloomingdale, Illinois. Fein suffered a professional crisis when 10 of his clients abruptly canceled their contracts, choosing ChatGPT over his services. Even though one of the clients later rehired him, his income was no longer enough to support his family, so Fein decided to seek refuge in a profession immune to AI's reach and enrolled in HVAC technician courses.[2] This all unfolded within eight months of the public release of ChatGPT. If we allow ourselves to become too wrapped up in our job titles, we will be hit the hardest, not just professionally but personally. We will close our window of opportunity by conceding defeat too soon.

We tend to forget that each disruption, minor or major, holds significance. It can propel us toward becoming the person we aspire to be if we let it, even though it may initially strip us of our current perceived self. But this journey isn't solely about navigating through the implications of AI; it's about reconstructing your identity, reaffirming your value, and acknowledging the multitude of experiences that have led you to this point in time and are preparing you for tomorrow. Neither the death of a loved one nor losing your job is required to experience these emotions. I've witnessed it in dozens of conversations with people proactively assessing their future and whether they are studying for roles that could become obsolete before graduation.

Facing irrelevance or starting anew after a job loss resonates with sensations associated with grief. It peels back layers of your identity that no longer serve you and compels you to realign your purpose with the present and reorient your focus. In our current time, attempting to cling to a rigid sense of purpose or a fixed career trajectory seems almost pointless; many of us will invariably face unanticipated deviations that carry us far from our original path. Consequently, our sense of purpose and worth must evolve to focus on versatile skills and emotions applicable to multiple roles, which provide a sense of certainty and grounding. My father's passing taught me the value of embracing three key qualities—adventure, connection, and growth—rather than solely relying on a specific purpose or vocation, while trying to acquire new skills during each personal evolution. Focusing on these qualities makes happiness and fulfillment attainable, independent of occupational changes or advances in AI.

EMBRACE RECONSTRUCTION

In the future, relying on a single career or ability might be insufficient, as versatility becomes paramount. This change demands the acquisition of numerous transferable skills suitable for a broad range of roles or entrepreneurial endeavors. In essence, we cannot outperform our current concept of self. Our present understanding of ourselves constrains us, keeping us entwined with our previous accomplishments, achievements, and traumas as well as our perceptions of our existing abilities and limitations. Still, the future doesn't just require you to adapt; it loudly and aggressively calls you to grow. If you fail to listen, it will scream louder until you do. Which brings us to the final three stages of grief.

The fifth phase, known as the upward turn, signifies a time when the turmoil of anger, guilt, or pain has subsided. You understand that change is a given, and you will need to be flexible to thrive. The sixth and most crucial phase is one of reconstruction. You start creatively piecing together how AI fits into your future and who you are relative to it. As for the seventh phase, we'll pull back the curtain on it when the time is right.

Two years after my father's death, I first delved into personal branding during my sixth phase, reconstruction, when I was seeking identity and purpose. As the internet expanded the global marketplace, I saw the significance of crafting a distinctive personal brand in the attention economy. This led me to formulate my unique interpretation of personal branding in 2009: self-expression amplified to influence and command attention. I kept this brand versatile to suit professional environments while maintaining flexibility, making it strikingly relevant today. That kind of brand is poised to become even more vital during the rise of AI as people crave more genuine human connections.

This idea takes on a new hue in the context of AI. CarynAI, an AI datebot released in May 2023, saw a 500 percent increase in usage after being featured in a *Fortune* article and gaining widespread attention.

The creation of Snapchat influencer Caryn Marjorie, the bot generated more than $71K in revenue in the first week after its beta test in 2023. The bot is the result of Marjorie's collaboration with deep fake startup Forever Voices, involving intensive training to accurately mimic Marjorie's mannerisms from her YouTube content and amplify her influence and self-expression. As Marjorie shared on Twitter, she has uploaded more than 2,000 hours of her content to become the first creator to be transformed into an AI.[3] Marjorie has found a way to duplicate her personal brand to engage broader audiences at scale by being unapologetically her. However, expanding to the masses may not be necessary for everyone.

Apple envisions individual AI assistants for each of our computers, which will supercharge our capacity to work more efficiently by optimizing the flow of information on our devices. Yet, what if we took it a step further, like Marjorie has, by cloning ourselves and developing AI assistants based on our knowledge, persona, and communication styles? Companies could then hire these AI clones to perform tasks we are currently handling, thereby creating new revenue opportunities. Imagine an online marketplace similar to Fiverr, where you can gain access to the avatars of top-tier business consultants, therapists, graphic design artists, copywriters, or your favorite author or influencer. Interact with these experts on demand, having meaningful dialogues with them at a fraction of the cost it would be to hire them. As training these models becomes more accessible, this isn't entirely out of the question, as Forever Voices has demonstrated. However, it's crucial to consider why people would choose you or your AI clone over others.

Whether online or offline, achieving professional longevity hinges on developing a personal brand that resonates with leaders and consumers alike. It can help you find new job opportunities, attain industry recognition, attract new clients, or even ensure that companies prefer your services to AI alternatives. Upskilling is critical in uncertain times, as we discussed in Chapter 5. And while AI may not take your job today, as it did Eric Fein's, that doesn't mean you shouldn't be

EMBRACE RECONSTRUCTION

prepared to meet the new demands of the attention economy. If the worst-case scenario doesn't materialize, you stand to gain the rewards regardless. This shift calls for today's workforce to acquire new aptitudes and adapt smoothly to emerging roles, occupations, and consumer demands. Key to this adaptation, beyond AI literacy, are transferable skills, which remain relevant and beneficial across any job, industry, or sector and constitute the foundation of your brand. They can range from specialized, hard skills like programming or data analysis to softer, interpersonal capabilities such as problem solving and nurturing relationships. Are you ready to evolve and bury your old sense of self to remain relevant?

As we contemplate our readiness for this seismic shift, let's delve into a pertinent study. Microsoft's 2023 report "Work Trend Index," subtitled "Will AI Fix Work?" suggests that companies that proactively adopt AI technology will disrupt the status quo, boosting both creativity and productivity among their employees. The company conducted a comprehensive survey of 31,000 individuals across 31 nations and analyzed numerous Microsoft 365 productivity indicators and employment trends drawn from LinkedIn's Economic Graph. The report identified the top five obstacles to productivity as follows:

1. Inefficient business meetings
2. Lack of clear goals
3. An overload of meetings
4. Lack of inspiration
5. Difficulty in finding necessary information

These obstacles can now be solved in part with AI as your copilot. However, the report also emphasized that collaborative skills with AI will be essential. These skills include analytical judgment, flexibility, emotional intelligence, intellectual curiosity, bias detection, management, and AI delegation (i.e., prompting an AI to produce work, such as writing a marketing email, rewriting a sales proposal, assessing legal documents, and so on).[4]

In light of the rise of AI and the need for collaboration, coupled with evolving consumer needs and more frequent career changes, both businesses and individuals must display agility. Surrounded by a rapidly changing business landscape, frequent personal rebranding becomes essential for maintaining relevance. In other words, effectively using and communicating with AI should be part of your personal and professional brand, reinforcing your identity as adaptable.

Instead of imagining a clearly defined purpose, envision the top three emotions you would like to evoke daily, weekly, or monthly, which will form the foundation of your brand and purpose. Reflect on your victories and hardships, identifying where you've shown grit, flexibility, emotional intelligence, and leadership. What key characteristics and traits define you—or could define you? Perhaps you're "solution oriented," "motivated," an unstoppable "problem solver," or quick to adopt new technology. Maybe you're a beacon of change tolerance in a tempest of uncertainty.

In an era where information is democratized, what makes your information invaluable to others? How can you remain relevant through continual learning and growth? As AI reaches further into our lives, what can you bring to the table that technology cannot replicate? It's time to dig deep and identify those unique traits, unabashedly embrace them, and couple them with AI to create a powerful synergy. Follow that by amplifying them throughout your social media channels to establish credibility. After all, you and AI don't just coexist; despite its threats and challenges, you scale together. But this will require something from you: reconceptualizing who you are.

AI's neural networks are continuously advancing and tackling more complex tasks. Thankfully, we can recalibrate our brains' neural networks through meditation and other techniques, enabling us to imagine and actualize our performance beyond the current perception of self—a vitally important skill when venturing into uncharted territories. The creation of new neural pathways and the reorganization of existing ones is an ongoing process. Yet we keep thinking the same thoughts and consuming the same type of content, reinforcing who we are and preventing us from moving beyond our perceived capabilities. This is something AI does not do, which is why it requires more from us. While meditation is an effective grounding force, I believe there is another one that is even more powerful.

Death and change, akin to both flow and states of hyper- or hypoarousal, bear a striking resemblance in their ability to manipulate our perception of time. Rather than accelerating or decelerating our perception of time, however, they pause it, effectively hindering our ability to see past the current moment and into the next. A smorgasbord of choices with no clear path can do the same, unless you can manipulate time yourself. This manipulation of time, in moments as life changing as a loved one's passing or the upheaval caused by breakthroughs in AI, becomes a peculiar tool, one that I, three months after my father's passing, began to finally understand.

Three months have now passed since Dad's death. I remove my framed goals from the wall of my home office in Melbourne, Australia that were penned during the peaceful early hours beside his bed. As I read them repeatedly, an intricate tapestry of "what-ifs" begins to unfold. Each read-through weaves a new thread, forming images of a life that could yet be mine if I manage to fulfill these ambitions. The "how" isn't my concern—I am content to wander in the vast realm of possibility for a moment.

Since I was young, I have always been fascinated with how to control my thoughts. When I was a life coach in my early 20s, I delved into hypnosis and neurolinguistic programming and studied how athletes prepare for a race through a technique known as mental rehearsal.

In mental rehearsal, you envision yourself practicing and mastering a task in your mind. Consider Natan Sharansky, a computer expert imprisoned for nine years in the Soviet Union on charges of espionage for the U.S. In solitary confinement, he would engage in imaginary chess matches with himself, famously quipping, "I might as well use the opportunity to become the world champion!" After his release, his mastery of the game led him to triumph over the reigning world chess champion, Garry Kasparov, in an exhibition game in 1996.[5] From his prison cell, he reprogrammed his brain to perceive beyond physical and mental limitations, amplifying his neuroplasticity and adaptability in a highly challenging environment. He refused to let himself be defined as a prisoner and created a new blueprint for his brain to venture.

Mental rehearsal can fall under various names, including visualization and imaginal exposure. The latter is an evidence-based technique used in psychotherapy for desensitizing the patient from trauma. By repeatedly confronting intimidating scenarios in a calm and governed state and creating imagined scenarios in which we adapt, we cultivate the ability to respond in preferred ways. Here you are doing the work in your mind before you take it to the real world by tricking your subconscious into thinking you've already been there and done it before, which means there's nothing to fear and only focus left to fuel you.

Athletes, billionaires, and entrepreneurs alike use this technique of mentally simulating the outcomes they aim for—winning a race, amassing wealth, or building a successful business. One of its major strengths lies in its ability to ease the negative emotions that often trigger procrastination, thus removing an obstacle to your objectives. Furthermore, it promotes the quick acquisition of new skills: One study found that nursing students who engaged in a 25-minute mental rehearsal session performed better in acquiring surgical skills than counterparts who merely viewed a training video. The students who used mental rehearsal were also better able to alleviate psychological stress, bolster self-confidence, and augment learning abilities by minimizing their cognitive load.[6]

Cognitive load indicates how much information your working memory can manage and store at a single point in time. Lowering this cognitive load is vital, as it enables efficient integration of accumulated information and paves the way for future planning.

An intriguing aspect of all the previous examples is that each involves creating new neural links that lead toward the person they aspire to be. This is an active process of crafting their future self within their mind, enabling them to transition smoothly into their ideal selves when required, speeding up the grieving and reconstruction stages. This is where the real appeal of mental rehearsal comes to light. The human brain despises open loops. The allure of television series like *Stranger Things* lies in their knack for introducing fresh, unresolved plot lines in every episode, creating an insatiable hook for your brain, which craves resolution. Mental rehearsal works similarly, laying down markers that your brain is compelled to connect. For instance, if you've mentally rehearsed acquiring new transferable skills but haven't yet figured out how, your brain wants to bridge this gap and starts tirelessly seeking solutions until it finds a satisfactory answer. This process also triggers your reticular activating system (RAS), a part of our brain that highlights valuable information from our surroundings and makes connections. This creates massive amounts of tension between where you are and where you want to be and dramatically increases your motivation by inducing your motivation molecule, otherwise known as dopamine.

I consistently employed this method throughout my 20s and 30s and continue to do so in my 40s, using it as a potent tool to sharpen my focus, write seven books, and confidently address audiences of thousands without being overwhelmed by nerves. Each morning before the summit began, I applied this technique to organize all the information I had received. This approach bolstered my courage and, even though I was blissfully unaware of them at the time, readied me for developments that had already begun unfolding in the atmosphere above the Atlantic Ocean, hundreds of miles away.

As dusk approaches on my first day at the summit, I can feel my vigor and enthusiasm slowly resurfacing. Even though I am nervous about meeting the media tomorrow, I feel optimistic, and I plan to keep that promise to myself and my Dad that I had made years ago. In the fading light, I decide to revisit the library on the way back to my Airbnb. Making my way past the lion statue known as 'Patience,' I find myself standing before his counterpart, 'Fortitude.' This embodiment of mental resilience was the other virtue that Mayor La Guardia deemed essential for the city to withstand the Great Depression. Rooted in the Latin word *fortis*, or "strong," fortitude is the strength of mind that enables a person to meet danger or bear pain or hardship with courage.

Yet this wasn't the first time I had crossed paths with fortitude. We had met years earlier, in the solemn quiet of dawn beside my father's bed. All my life, I'd seen this steadfast spirit personified in those around me, but it had never occurred to me that it could reside within me, too. "Maybe it's time for me to embrace and acknowledge my own fortitude," I wonder. I have overcome great odds to be here, and only fortitude will see me through the next stage.

CHAPTER 6

ADVENTURER'S HANDBOOK

Rule #6: Embrace Reconstruction

In the face of rising AI competition, forging your distinctive personal brand requires authenticity, creativity, agility, and a clear comprehension of your emotional responses to change, especially under the threat of irrelevance. Understanding your current position in the stages of grief—denial, pain, anger, depression, and so on—can guide the emotional transitions linked to your personal reinvention. Consider the following elements to begin the reconstruction stage.

Pivot and Prosper: The Art of Personal Reinvention in an Automated Era

- **Authenticity**: How do your true self and personal experiences uniquely shape your brand to connect deeply with your audience in a way that is distinctly different from AI-generated content?
- **Emotional awareness**: How do your emotions influence your response to change? Acknowledging where you currently stand in the grieving process can give you insight into how to progress toward acceptance and constructive action.
- **Creativity**: How can you weave your innovative narratives into your brand, creating layers of complexity that set you apart from commonplace AI-generated content?
- **Resilience and agility**: What strategies can you employ to stay relevant and distinguish yourself as AI continues to proliferate?
- **Storytelling**: How can you shape significant changes and challenges from your experience into compelling narratives that amplify the human connection with your audience beyond the capabilities of AI?
- **Brand consistency**: How are your brand's tone, design, and messaging reflective of your true self, and what sets it apart in an AI-dominated landscape?
- **Personal growth**: How do the experiences derived from navigating through the stages of grief manifest into personal growth, resilience, and authenticity that fortify your brand against AI competition?

Visualization/Mental Rehearsal

- **Mentally prepare for your future**: Practice mental rehearsal to sharpen your skills and overcome limitations, drawing from the example of Natan Sharansky, who mastered mental exercises in solitary confinement. By envisioning yourself stepping into this reimagined version of yourself, you can forge new neural links and increase your adaptability. Find this specific genre of visualization at areyouunstoppable.com.

Decoding the Top 10 Threats of Artificial Intelligence

Navigating the ninth threat on our list is irrelevance. The relentless progression of AI wields the potential to eclipse human capacities. The need for human involvement dwindles as AI seizes the reins on tasks and processes, sometimes even outshining us in areas requiring abstract thinking and complex analysis.

9. **Irrelevance:** Confronted with an AI-driven societal makeover, the quest for personal identity and purpose will likely be our hardest battle, eliciting our deepest fears about irrelevance. This shift could engender an existential crisis as traditional human roles become obsolete. With AI reshaping the social fabric, professions, and even personal identities, we may grapple with finding purpose and meaning in this evolving landscape.

Rule #7

Find Your Frequency

DECKED OUT IN A TAILORED BLAZER, a spotless white shirt, jeans, and brown Italian shoes, I stroll casually into the spectacular entrance of The Roosevelt Hotel. My gaze is magnetically pulled to the lobby's breathtakingly high ceiling, a beautiful dome cradling elements of Art Deco design, carefully crafted moldings, and gold-trimmed details that speak truth to its iconic era. Like a time machine springing to life, I'm hurled back into the heart of the exhilarating Roaring Twenties.

This era was characterized by a thriving economy, swift cultural and societal changes, and an atmosphere sizzling with energy. In September 1924, just a few years before the Great Depression hit in 1929, the magnificent hotel opened its prestigious doors. Dedicated to America's 26th president, Theodore Roosevelt,[1] and nestled in Midtown Manhattan, the hotel attracted a crowd of aristocrats, celebrities, and socialites. As I stroll through the lobby, I can almost sense the lingering spirits of the past.

Today, on the second day of the summit and the first day of officially meeting the media, I take the Art Deco lift up to the conference room and sit down, tired yet determined. In an adjacent room, esteemed TV producers, magazine editors, and publishers stand prepared for the onslaught of pitches from 100 eager attendees.

"Let the hunger games begin!" I announce to a new friend. As they wish me well, I step into the room. The pressure descends on my shoulders like a palpable weight. Across the room, an editor from *Marie Claire* presides over a queue of 13 keen professionals, each bracing for their allotted 60 seconds to pitch themselves as experts within their field. They're waiting for a lucky break - be it media exposure or a book deal. For me, it is both. Nearby, a globally renowned publisher, towering 6 feet tall, exudes an aura of no-nonsense professionalism. His Brooklyn accent rings from across the room. Despite not being easily daunted, he rattles me. Doubts begin to set in as I ask myself, "Have I made the right decision travelling 10,000 miles to be here?"

As my gaze darts nervously across the room, I gravitate towards Cheryl Wills, a television presenter from NY1. During our conversation, I pitch my expertise in personal branding, making sure she knows I'm only a call away if she ever needs to interview an expert on live TV. With a massive smile, she kindly inquires about my origins and congratulates me on making the journey. To my surprise, she hugs me as we say our farewells. Perhaps she senses I need reassurance. Unbeknownst to Cheryl, she is poised to play a crucial part in the events that will bring the bustling city to a halt in less than 72 hours. Yet, as I walk away, I begin to sense an inner conflict stirring. My focus is not aligned with my energy levels; it is scattered. Here, I stand in a room full of once-in-a-lifetime opportunities, yet much like the traffic in Times Square during rush hour, my thoughts dart haphazardly in every direction.

I know I must resist the urge to sleep, but the pressure is mounting. It is 9 a.m. here, but it is 1 a.m. the next day in Australia, and my internal clock is in chaos; my mind racing ahead to the future while I remain tethered to the past. At that moment, someone approaches me and exclaims, "Ben, you look as pale as a ghost!" My mental acuity is compromised, however, I continue undeterred. I'm casting my net wide, making pitch after pitch to all media associates, except for the publishers. My presentations are disjointed, and I'm not achieving the desired impact.

Yet, isn't this the usual state of affairs when we dive headfirst into the deep waters of the unknown? Getting to the bottom of such sensations is instrumental to our onward progression.

Often, we perceive the essence of freedom as the ability to make choices—from monumental life decisions to the mundane selection of dinner. But the question arises: Can an overabundance of options end up being paralyzing? As we confront infinite possibilities and pivotal existential questions presented by the era of AI, like determining our career trajectories or deciding on the ethical boundaries of technology, we risk falling into a paradoxical trap called decision paralysis, driven by an overload of choices.

Choice overload, also known as overchoice, creates a cognitive bottleneck when faced with an overwhelming array of options, impeding our ability to make sound decisions. This not only compromises the quality of decision-making but also diminishes contentment with the ultimate choice. As we navigate challenging and unknown territories beyond our current understanding, there's a risk of losing touch with our innate ability to make well-informed decisions that include foresight.

In 2023, the rise of 'wrapper' startups fueled by ChatGPT's technology exemplified this dilemma. Eager to secure a first-mover advantage, these startups added value by incorporating existing application programming interfaces (APIs), such as ChatGPT and Bard. However, they failed to anticipate future trends that could render their services obsolete. OpenAI's modifications, particularly enabling ChatGPT users to work directly with PDFs, posed a direct threat to the 'wrapper' startups, challenging their newly formed business models.

This example illustrates how the rapid evolution of AI technology can contribute to the degradation of strategic decision-making. Navigating this constant change requires the right mindset and the ability to anticipate technological advancements to avoid pitfalls in creating businesses and careers that may falter due to shifting trends. Opting for short-term decisions may seem lucrative initially but can lead to costly repercussions as industries and technologies advance, potentially devaluing or eliminating short-lived gains.

In such scenarios, the concept of rhythm can be an antidote. The Roosevelt Hotel's rich history and the surge of profound emotions I felt there served as the catalyst for this understanding. The same rhythm—reflected in the rapid tempo and energetic music that once filled the hotel during the Roaring Twenties—can be applied to our decision-making process. Just as the harmony of bustling energy and cadence was essential to the hotel's atmosphere, life's decisions too require a rhythm—a balance between seizing the right opportunities and avoiding the wrong ones.

By aligning decision-making efforts with our natural rhythm, individuals can optimize their energy, enhance cognitive performance, prevent decision fatigue, adapt to change, and strike a harmonious balance between focus and recovery. This alignment contributes to an overall framework for making more sound and effective decisions about your future.

Ultradian rhythms—the biological rhythms of your body—are the unseen orchestrators of our mental, physical, and emotional well-being. Similar to the constant lapping of waves on a beach (brain waves) or the steady ticking of a heart (cardiac rhythms), ultradian rhythms are physiological patterns that can be measured, the hidden metronome to which our bodies synchronize to function correctly. They are inherited rhythmic blueprints inscribed within your DNA, a product of your "clock genes." These genes conduct a symphony of bodily functions in exquisite temporal harmony, ensuring everything operates in sync with the passage of time.[2]

Seeking to understand these rhythms leads us to a key figure who has unraveled their secrets and enhanced our understanding of the sleep and wake cycles. Emerging amid the Roaring Twenties, Nathaniel Kleitman, initially penniless upon his arrival in New York City, would later signify an era of immense expansion in our knowledge of these patterns.

The 20-year-old Kleitman started with nothing but ambition. He dedicated himself to his education at the City College of New York and continued his journey at the University of Chicago's Department of Physiology, where he earned a PhD. His groundbreaking dissertation, titled "Studies on the Physiology of Sleep,"[3] laid the foundation for his career as the first academic scholar to focus exclusively on sleep and secured his place as the pioneer of sleep research.[4] Among his many contributions, Kleitman identified a pattern known as the "basic rest-activity cycle," which was pivotal for numerous foundational discoveries, including circadian rhythms.[5] A two-week investigation aboard a submarine in 1948 led to remarkable insights about fluctuations in mental sharpness, underscoring the complications of adjusting to changing time schedules and the importance of considering the body's innate rhythms when planning workers' schedules.[6]

Kleitman's discoveries also included the observation that our bodies adhere to a repeated 90-minute rhythm characterized by shifts from states of higher alertness to lower alertness, which was later dubbed the "ultradian rhythm" by fellow researchers. Rhythm refers to the ebb and flow of these wave patterns, and ultradian denotes multiple occurrences within a day. They are principally designated to orchestrate a three-stage cycle, from manufacturing energy; second, its effective utilization; and third, recovery (our healing response). While AI operates effortlessly on the first two stages, it disregards the third, hence exempting itself from fatigue or the need for rest. Such an exemption, however, bestows AI a competitive advantage, making it difficult for humans, who necessitate all three stages for optimal focus and decision-making, to compete on an equivalent level.

AI, just like your brain, manages the ability to concentrate by prioritizing critical information and promoting it for in-depth processing, all while keeping disruptions from nonessential fragments at bay. This operation is often referred to as efficient selection. But this efficiency can dwindle in us when we disregard Kleitman's "basic rest-activity cycle," giving rise to what is known as decision fatigue, a state of mental saturation that can hamper our capacity for intelligent decision making.

In contemporary society, we frequently disregard Kleitman's discoveries. We inundate our systems with caffeine to keep going, turn to incessant social media scrolling when our minds require rest, and constantly push ourselves to exceed our previous efforts, interrupting intrinsic rhythms that are crucial for adaptability and focus.

In a telling case of how ultradian rhythms function, Anders Ericsson, a psychologist from Florida State University, conducted a 1993 study to identify the factors contributing to achievement. His study, "The Role of Deliberate Practice in the Acquisition of Expert Performance," revealed shared traits among successful violinists. Notably, they split their practice into three daily sessions. Each session lasted roughly 90 minutes, punctuated by 20 minutes of rest. Ericsson emphasized that the key lies not only in the quantity of practice, but also in the quality of practice.[7] In this scenario, willpower wasn't praised; it was frowned upon.

Our hustle culture glorifies willpower, but we quickly forget that it is about exerting control over critical aspects of our lives, whether external (factors beyond our control, such as certain AI developments) or internal (energy regulation and the battle against mental exhaustion in comprehending what it all means). The invocation of this control suggests that the situation is out of control. Relying solely on willpower to utilize energy proves ineffective as it means we're engaged in a constant struggle against friction. By synchronizing with our ultradian rhythm and averting an escalation into states of anxiety, fatigue, and burnout, we lessen our dependence on willpower, freeing up our mental capacity to confront the seemingly insurmountable developments that are taking place. It also fosters habit formation. How? It reduces friction.

If the goal is to adapt by choosing our next steps carefully, we must ask ourselves how we can make the process easier. The solution is to adopt habits that align with our purpose and our rhythms. Obstacles that prevent you from accomplishing your goals create friction: It is a force resisting the movement of one object against another. Friction manifests in three ways: distance, time, and effort. Yet we must upgrade our understanding of this within the context of AI. "Distance" refers to the knowledge you must learn. "Time" relates to the rapid changing of events, the time it will take to adapt, and the manipulation of time to achieve flow states. "Effort" pertains to using your mental and physical reserves to discover and actualize new opportunities.

The forces that cause friction accumulate over time, like sleep debt, overchoice, and decision fatigue. You may have encountered this friction in the previous chapter, where the struggle between your past and future selves for relevance was evident. In this case, friction has a purpose: to ignite a fire of change within you. Even if it is wildly uncomfortable to sit with, there are times in our lives when it is necessary.

Integrating AI into the workplace has the potential to alleviate friction and bolster productivity, even if it's not always so straightforward. As you are now aware, it can also generate friction, particularly when striving for a first-mover advantage and capitalizing on emerging opportunities while navigating new threats. But by reducing cognitive friction and combining AI, a better environment is formed for achieving a flow state, where we can assess our options carefully and reduce our need for willpower.

AI technologies can support us in each stage of Kleitman's model — energy creation, utilization, and recovery — and can assist individuals in managing stress, boosting productivity, and enhancing overall well-being. In the energy creation phase, AI tools can enhance nutrition by tracking intake and making personalized recommendations, promote exercise via tailor-made workout routines, and assess and improve sleep quality via intricate pattern analysis. During energy utilization, AI aids in time and task management by optimizing schedules, heightens focus and productivity via personalized strategies, and nurtures emotional intelligence through self-awareness exercises and stress reduction techniques. In the recovery phase, it monitors stress levels via wearable devices, recommends personalized relaxation techniques, and encourages regular breaks and downtime. Utilizing AI in these ways not only helps prevent burnout and improve personal well-being;

it's a critical tool in maintaining and enhancing our decision-making capabilities by freeing up mental bandwidth for more important work.

However, there exists another unseen force, one that surprises as it deftly winds together all three of Kleitman's prescribed steps, not only tempering friction, but also ingeniously fostering momentum. This force also demonstrates that great decisions address the root cause, not just the symptoms, and we can find it just a few blocks away from the Roosevelt Hotel.

Often, when people think of the Rockefeller Center at 45 Rockefeller Plaza, the annual Christmas tree lighting up the night sky comes to mind. The tradition dates back to 1931, when a 20-foot balsam fir was decorated with homemade garlands. Workers at the center pooled their money to buy and place the tree as a symbol of hope during those dark days. A photo online vividly captures the first tree's presence. It's a riveting sight: laborers in overalls standing amid heaps of debris, with the tree standing out in stark contrast to its surroundings. At the time, the workers were unaware that their actions would help build enough momentum to help New York emerge from the Depression. Simultaneously, they laid the foundation for an enduring annual tradition that is still celebrated by millions to this day.[8]

In the intricately woven history of the center, innovation, poverty, and wealth play significant roles. The project's plans were solidified one day before the 1929 stock market crash.[9] During a time when more than 750,000 New Yorkers were unemployed, the Rockefeller Center emerged as a beacon of hope. Built between 1929 and 1940, the Art Deco-style project ignited a surge of momentum.[10] An estimated 40,000 to 60,000 jobs were created during its construction alone. Raymond Fosdick, who led the Rockefeller Foundation, once estimated that the number of people involved reached an astounding 225,000, counting the on-site laborers and those who manufactured the materials for the complex.

THE WOLF IS AT THE DOOR

The Rockefeller Center underscores the lasting impact of strategic foresight, emphasizing that even in the face of daunting odds, well-calculated decisions can propel both resilience and prosperity.

Now, fortified by this knowledge of collective determination, let's take our exploration to new heights. Close your eyes and imagine as we lift off the ground, zooming straight up through 70 stories to discover the breathtaking view from the Top of the Rock's observation deck.

Stepping onto the platform, the view sweeps across an awe-inspiring panorama. Central Park, a sprawling green oasis embraced by the city's high-rises, rolls out on one side. On turning, the eye runs along Manhattan's backbone, resting upon a testament to resilience: the World Trade Center. Intriguingly, there's another presence here that eludes sight. New York pulsates with an undeniable rhythm, an urban heartbeat, if you will, but it also hums at a unique frequency, a signature wavelength all its own.

Just like New York, we have our own distinct frequencies in the form of brain waves, which can be thought of as the unspoken language of our minds. They were first found in animals around 1875. The groundbreaking documentation of these waves in humans was by Hans Berger in 1925, who hoped to show that the human brain's electromagnetic fields could serve as a medium for telepathy.[11] Although he was unsuccessful in this endeavor, his findings further emphasized that our world, like our brains, is a complex interplay of rhythms and frequencies that can be measured, just like our ultradian rhythms. His work contributed to the wide adoption of EEG devices.

The varying frequencies of brain waves signify different states of consciousness or alertness. These waves are formed by neurons in the brain that generate electrical charges during their interactions. Meditation can shift them from a higher frequency to a lower one, altering your emotional state. When beta waves (around 13 Hz and higher) dominate, they indicate a state involving active thinking and self-awareness. In this state, a person is highly attuned to the outside world.

When they are disengaged from external surroundings, delta waves (0-4 Hz) associated with sleep, predominantly appear. When theta waves (4-8 Hz) are most prominent, a person focuses primarily on their inner world. This is known as the theta state, a realm of vivid mental imagery. Some people undergoing alpha-theta therapy have reported experiencing an "inner healer" who embodies their transition from feeling powerless to empowered. Alpha brain waves (8-13 Hz) can be viewed as a bridge between the external and internal worlds, allowing for a seamless transition between the different states of consciousness.[12] Finally, there are gamma waves (25-140 Hz). These brain waves are associated with extensive brain network activity and are thought to be involved in active information processing, learning, and memory functions.[13]

The importance of matching your frequency to the activity—whether it's sleep, relaxation, intense focus, or making decisions about your future—cannot be overstated. However, we often encounter scenarios where our timing aligns with the tasks at hand, but our frequency does not. It's akin to tuning an old-school radio to find our favorite station. The objective is to sync with the frequencies that boost productivity, reduce resistance, and foster renewal, but it's common to get lost in the static between stations and become overwhelmed by options. Of course, as with everything, there is a certain subtlety to understanding brain waves.

Delta waves, the slowest type, are common in infants and young kids. These waves are instrumental in attaining the utmost levels of relaxation and the most recuperative stages of sleep. In cases of brain damage, cognitive difficulties, learning problems, and severe ADHD, delta waves seem to be more evident. Insufficient delta wave production at night hinders the body and the brain from rejuvenating themselves. Ensuring adequate delta waves during sleep promotes natural healing and deep restorative rest, enhances immune system function, and nurtures new brain connections.[14] This is key, as not all sleep is created equal, and not every problem can be solved with the same brain wave frequency.

During deep sleep (delta), our brain waste removal system (BWRS) activates, expelling toxins and waste from our central nervous system.[15] Fascinatingly, it was Aristotle who first came up with the idea that sleep triggers brain cleansing; it wasn't for another 2,000 years that this was confirmed by multiple studies.[16] The accumulation of toxins in the brain has been linked to conditions such as Parkinson's disease, brain tumors, trauma, Alzheimer's,[17] and neuroinflammation, which is associated with brain fog.[18] Scientists have also discovered that neuroinflammation is a potential biomarker in anxiety disorders.[19] Furthermore, the activation of the BWRS during sleep has given rise to a new idea that is the subject of much discussion: nighttime stimulation of the BWRS might offer a groundbreaking approach in neurological rehabilitation.[20]

During sustained mental concentration, your body and brain consume vast amounts of energy. Around the 90-minute mark, you hit your productivity peak, called the "ultradian performance peak." Meanwhile, toxins from your environment and mental and physical effort build up within your system. Over the next 20 minutes, this buildup manifests as fatigue, causing your performance and productivity to decline as you enter a low-energy phase known as the "ultradian trough."[21] Therefore, it is crucial to make decisions surrounding your future during your productivity peak and avoid doing so during your trough. Making decisions when experiencing decision fatigue can lead to poor outcomes and choices you may later regret. If we fail to take frequent breaks or get enough restorative sleep, toxins accumulate causing a drop in mental performance and an increase in impulsivity.

Interestingly, frequency can also affect brain waves. For example, you can stimulate the production of certain brain waves—especially gamma waves—by using sound therapies like binaural beats or flickering light therapy.[22] Binaural beats are a fascinating auditory trick in which two distinct tones with varying frequencies are played in each ear. This discrepancy creates the illusion of a third rhythmic beat, prompting neurons across the brain to emit electrical signals syncing up with it.[23] Although studies on the impacts of binaural beats are relatively limited, their findings support the idea that they offer health benefits, especially in areas relating to anxiety, mood, and performance. One 2023 study revealed that binaural beats set at 40 Hz enhanced focus, information processing, and memory retention.[24] Early research also revealed that the 3 Hz frequency offered sleep-enhancing benefits by stimulating delta brain activity. As a result, the duration of the third sleep stage, during which the BWRS is activated, was extended.[25]

Our brain frequencies can also be influenced through the use of pulsed electromagnetic field (PEMF), a technique that intriguingly simulates the Earth's natural electromagnetic field. The ever-changing frequencies intrinsic to this field are known as the Schumann Resonance, named after German physicist Winfried Otto Schumann. Spanning an array of 1 to 20 Hz, they peak significantly around 7.83 Hz.[26] Intriguingly, these frequencies find a parallel in our lives. Just as we manipulate and utilize pulsed PEMF technology to influence our brain frequencies, we're undeniably intertwined with the natural electromagnetic rhythm of our planet.

PEMF therapy, which involves subjecting the body to electromagnetic fields, can trigger various biological processes. In recent years, a resurgence in PEMF's popularity is largely due to advances in wearable technology. These modern devices, with mechanisms designed to generate pulsed electromagnetic fields, have become more accessible, affordable, and user-friendly. They can be comfortably worn or placed on the body, offering the advantages of PEMF therapy from the comfort of one's home. These wearable devices not only empower individuals to access their own healing abilities but also facilitate the integration of this treatment into standard health and wellness routines.

PEMF operates by emitting specific frequencies, such as theta and delta waves, that resonate with the brain's natural frequencies. These waves can induce desirable states like relaxation, focus, and sleep. By modulating brainwave patterns, PEMF devices stimulate the body's innate healing mechanisms and provide various health benefits.

Where PEMF shines is in its ability to module the release of cytokine during inflammation stages, improving human and animal tissue recovery.[27] Another study showed promising results in stroke rehabilitation and recovery, as patients in the PEMF group exhibited approximately 45 percent improved depression scores and a 35 percent increase in cognitive function.[28] For over three years, I have reliably turned to a PEMF device as a tool to mitigate inflammation and cultivate a state of tranquility and focus to ensure I match my frequency and rhythm to the task at hand.

By harnessing the restorative abilities of our brains, we can effectively manage our stress responses. This management, achieved through harnessing frequency, momentum, and rhythm, equips us with the basic tools required to make well-considered decisions. Such decisions, made from a well-balanced state of mind, can help us avoid future regret often born from acting on impulse, fatigue, or fear. It's essential when choosing AI tools to aid us, that they reflect the three stages outlined in Kleitman's model: energy creation, utilization, and recovery. Together, these elements form a cycle which exceeds the scope of Kleitman's original concept of the basic rest-activity cycle. It positions us for success in a new uncertain world, one that would have felt familiar to those workers who raised the first Christmas tree at Rockefeller Center. But when we disregard the recovery dimension, we sever our connection to an invaluable asset within each of us: knowledge. With the frequency-charged experience at the Rockefeller Center and the rhythmic echoes of the roaring twenties at the Roosevelt Hotel in the backdrop, we rewind our story, landing at the end of an intense day at the media summit.

Exhaling a breath of relief as I crawl into bed back at my Airbnb, I welcome the calm at the end of the long day. Though I had practiced my pitches diligently, lack of sleep had eroded my focus, standing between me and the many opportunities that lay before me. I now recognize that being in the wrong mindset and being overwhelmed by options could cost me dearly. Like a light switch being flicked off, I forgot who I was and what I was truly capable of.

As sleep beckons, my mind circles around a fascinating revelation I have recently encountered, wondering if it will seep into my dreams tonight. Six feet under Bryant Park, next to the New York Public Library, a clandestine passage is buried. This 120-foot subterranean tunnel serves as an unnoticed lifeline, linking the library's hidden storage vault to the main branch and sheltering an awe-inspiring collection of nearly 1.5 million books.

This vast horde of knowledge, hidden from view, remains uncharted territory for most, its labyrinth of secrets revealed only to a privileged few. It makes me wonder whether I possess all the keys to unlock all the knowledge I have stashed in my subconscious over the years. Have I lost them, or are they tucked away in some forgotten corner of my mind? It seems plausible that the treasures of opportunity and progress are still present—they are just ignored. I just need to get in the right rhythm and frequency to perceive them.

A vital realization dawns upon me: Knowledge, applied correctly and timely, forms a crucial element to effective decision-making. Because, without knowledge, my decisions will be ill-conceived. If knowledge is the currency of the new age, then learning is the act of forging my own keys to unlock the abundant wealth I already possess. I can tap into the frequency of change and find my flow by reducing friction. For that, I will require another key, one that has been misplaced by many — including you.

CHAPTER 7

ADVENTURER'S HANDBOOK

Rule #7: Find Your Frequency

Creating an "inner symphony" that combines binaural beats, frequency, AI, meditation, and the ultradian rhythm can help optimize yourself for a calm, focused state of flow, which is crucial for adapting to a fast-paced environment. Here's a strategic framework to "find your ideal frequency":

A Symphony of Self-Optimization Techniques

1. **Tuning In to Your High Productivity Frequency**
   - **Binaural beats**: Immerse yourself in binaural beats as part of your daily routine to foster deep relaxation and enhanced concentration for improved productivity.
   - **Meditation**: Carve out time for meditation sessions. Regular meditation can reduce stress and heighten focus, so make it an integral part of your daily regimen.
2. **Synchronizing with your ultradian rhythm**
   - **Manufacturing energy**: Adopt a balanced diet, get ample sleep, and exercise regularly. These key components help provide the required energy to keep your body functioning optimally.
   - **Effective utilization of energy**: Stay aware of how your energy ebbs and flows. Do your most demanding tasks during high productivity periods, which coincide with your ultradian rhythm cycles.
   - **Recovery**: Integrate recovery techniques into your routine. This helps revitalize your energy levels and maintain your productivity over the long haul.

3. **Activating your brain waste removal system (BWRS) through deep sleep**
   - **Sleep routine optimization**: Cultivate a consistent sleep schedule, which promotes deeper, more restful sleep.
   - **Sleep hygiene**: A conducive sleep environment, like a cool room, can induce better-quality sleep.
   - **Presleep wind-down**: Establish a calming routine before sleep to ease your transition into a restful night.
   - **Morning sunshine**: Soak up morning sunlight shortly after waking. This can positively impact your sleep cycle, energy levels, and overall mood.
   - **Optimizing the ultradian rhythm**: Recognize and align with your unique 90-minute cycles. These peak productivity periods help maximize focus and minimize energy drain.
4. **Turn friction into flow**
   - **AI integration**: Explore integrating AI technology into your lifestyle, which can effectively reduce friction and decrease decision fatigue.
   - **Identifying friction points**: Keep an eye out for any elements in your life that contribute to unnecessary friction. Address these friction points to streamline your routines and elevate overall productivity.

Finding your frequency by leveraging the combination of binaural beats, meditation, ultradian rhythms, and intelligent tech integration plays a pivotal role in achieving a flow state. Identifying this efficient and harmonious balance aids in navigating a rapidly evolving environment, optimizing your well-being, and enhancing your overall productivity.

RULE #8

Boost Your Brain Power

I DASH OUT of the five-story apartment, past Mayor La Guardia's statue, searching frantically for the bodega I had seen earlier in the day. I need water and canned goods desperately. Suddenly, a trash can goes flying across the once busy Manhattan Street. "Oh no, it's too late," I think as I sprint back to the apartment, soaked by the cold rain and buffeted by strong winds. Alerting Alex, I reveal that every store is closed. Our only option now is to fill the bathtub with water, batten the hatches, and hope we will be safe overnight. We will search for food tomorrow.

After wrapping up the summit, I had rendezvoused with Alex, a good friend from Australia, whose holiday in the Big Apple coincided with my trip. Since my time at the Airbnb had ended and all other accommodations in the city were occupied, leaving me without further lodging options, he offered to share his rental apartment on Bleecker Street, in the heart of Greenwich Village. Still in a daze from my whirlwind week at the media summit and still recovering from my extreme sleep deprivation brought on by jet lag, I had failed to keep an eye on the news. The very media people I had pitched to were now focusing their full attention on the storm of the century; my $15,000 investment in my future had just gone up in smoke. Along with countless New Yorkers, they were laying low as Hurricane Sandy swerved abruptly to the left and began moving squarely toward New York, catching millions off-guard. Its winds spanned an astonishing 1,000 miles—three times the size of a standard hurricane. Panic-stricken texts from friends and family back in Australia started flooding in as news networks back home began reporting on the unfolding crisis. A popular meme circulated on Facebook depicting the Statue of Liberty taking cover behind her pedestal. As I urgently turned on the TV to see what the panic was about, I saw a familiar face from the summit reporting on the situation—Cheryl Wills from NY1.

Growing up in Australia, Alex and I had both believed that Americans tended to exaggerate. Since we had never encountered a hurricane, we dismissed the forecasts as overstated and failed to prepare. We were horribly wrong!

As Sandy made landfall on October 29, 2012, ferocious winds battered the city, and the severity of the storm quickly became apparent. For the first time since the tragic events of 9/11, all bridges and tunnels connecting Manhattan to the mainland were closed. Rapid currents flowed through the lower Manhattan tunnels in the Financial District, transforming them into raging rivers. Water continued to inundate the streets, carrying along cars and yellow cabs and even encroaching on the 9/11 Memorial. A staggering 375,000 residents of New York City received mandatory evacuation orders.[1] As people rushed to evacuate the lower areas of the city, they discovered the subway system was already closed. Tensions increased with long lines at the bus stops. The subway was the city's lifeline, and the daily commute of millions was critical to the city's economy. Floodwaters engulfed it in a stunning spectacle, spilling out onto the streets just a block from our building.

A twisted construction crane hung precariously from a 75-story high-rise as powerful winds battered the structure. Fire department officials were forced to evacuate guests from a nearby hotel and cordon off a seven-block area—including the hotel where I stayed the first night of my visit to New York.

I alternated between watching the TV and looking out the window, my apprehension escalating with each passing moment as we observed the chaos unfold. Although we didn't know it at the time, at precisely 9 p.m., a ConEd substation in Alphabet City flooded, causing a massive transformer to explode about 20 blocks from us.

The explosion, visible as far away as Brooklyn, knocked out part of the island's grid, and we, like the rest of lower Manhattan, were plunged into darkness.[2]

Just as persevering through a citywide blackout requires effort and ingenuity to ensure the lights stay on, maintaining our cognitive lights—our mental endurance and inherent adaptability—is crucial to preserving our resilience in this rapidly changing world. With the demands of daily life already stretching many people to their limits, keeping pace with the monumental implications of AI can feel insurmountable. This revolutionary technology promises to reduce the cognitive burdens for millions, but it may come at the cost of rendering their skills redundant and disrupting their mental well-being. The potential benefits you'll reap from this revolution remain to be seen. However, your focus, the energy fueling your ability to adapt will ultimately determine your destiny.

The question looms large: Can we manage adding another task to our already packed agenda, namely learning to incorporate AI into our lives, careers, or business models in order to thrive? Seeking the answer to this question will take us on a journey beyond Sandy's devastation, fast-forwarding three years to a chilly winter in Adelaide, Australia, in 2015.

That year, an unexpected sequence of events unfolded. A heavy tide of exhaustion and melancholy abruptly cut my lights, catching me off guard. It was so severe that friends observed me slurring my words. As a result, I was afraid I might have to give up my passion: public speaking and writing. Even though my spirit was resolute, the limitations of my body, especially my brain, persisted. To find answers, I embarked on an ambitious mission with the little energy I had left to biohack my body and brain within 90 days. We surveyed more than 70,000 people internationally online to see if we could identify hidden patterns. The results were astonishing and revealed far more than we could have anticipated.

How? I examined peak performance as though it were a two-sided coin: biology on one face and psychology on the other. Many self-help experts concentrate on just one aspect while neglecting the other and failing to grasp how fundamental the role of energy creation is for psychological change to occur. By aggregating more than a million data pieces, we connected psychological and biological indications to performance, resulting in a novel model for realizing peak performance that has transformed the lives of tens of thousands globally.

We identified powerful correlations among crucial triggers, such as excessive caffeine intake, dietary habits, gut health, and issues like procrastination, fatigue, suicidal thoughts, adaptability, and focus. From the pool of more than 70,000 respondents, we discovered that a mere 6.1 percent could be classified as actual peak performers. What differentiated their behavior from the underperformers was that they displayed minimal biological symptoms, implying their brain and body had ample energetic resources available for optimal cognitive performance.

Unhindered by brain fog and fatigue, they achieved their goals 84 percent of the time, while those in depletion could only reach their goals a meager 8 percent of the time. They also demonstrated a robust ability to adapt swiftly to changing environments. What was their secret? Their potent ability to manufacture energy acted as a propelling force—this is the initial stage of Kleitman's model, which we discussed in Chapter 7.

Following Kleitman's second stage—using energy effectively—they efficiently channeled this energy and focused on accomplishing their goals. They were operating in their optimal arousal zone, not just for focus, but also for change. Yet the majority either overenergized themselves with caffeine, leading to anxiety that disrupted their focus, or succumbed to overwhelming fatigue, leaving them too drained to operate. They are at risk of being overwhelmed by AI, not just because their emotional regulation is poor, but also because their creation and use of energy is ineffective.

Like a hurricane functions as an engine, individuals typically demonstrate a similar pattern in life. They spring into action when struck with an initial wave of inspiration, using it as fuel, just as hurricanes draw energy from warm waters. In their quest to achieve their goals, they work tirelessly, pushing the edge of their capabilities. But once the fuel that maintains their momentum is exhausted, they diminish in strength and intensity, losing their drive and eventually fading out. They hit their peak too soon.

There are others who are ineffective at using the energy they have and expend their cognitive resources uselessly, sometimes in online quarrels with people they've never met on subjects that are none of their concern. What's even more detrimental is when individuals fall into a cycle of intense advancement followed by complete inactivity. This erratic progression hampers their ability to generate steady forward momentum in their careers and forces them to contend with cognitive and emotional friction.

Our survey paints a telling picture: over half the participants admitted that their ability to concentrate diminished as their biological symptoms escalated. There was a startling divide in the data: While only 8 percent of underperformers could sustain focus over prolonged periods, a whopping 85 percent of peak performers managed to do so. This is primarily attributed to whether they executed the third step in Kleitman's model—whether they expended all their energy without allowing ample time for recovery. However, the most consequential oversight was the neglect of Kleitman's first stage—manufacturing energy—which is vital for sustaining yourself through rapid and lasting change.

Our research revealed that a sizable chunk of mental stress comes from biological triggers, but it is wrongfully blamed on insufficient willpower. Underperformers attribute blame to themselves instead of addressing the core issue, perpetuating a whirlpool of self-condemnation. Acute psychological stress has been identified as leading to increased inflammation and a heightened focus on negative details (in other words, reprogramming your brain's attention system to seek out negative information or worst-case scenarios). This then exacerbates any underlying conditions, creating a vicious cycle. The ensuing rise in inflammation can trigger symptoms of depression, bringing about a drop in both physical and mental energy.[3]

When we zoom out to the broader landscape of change and how it relates to psychological and biological stress, it becomes clear that our current transformations are akin to those of the Industrial Revolution. Just as positive innovation then demanded a steady supply of energy, so does AI's transformative impact. In the early stages, the rate of job loss is anticipated to exceed job creation, requiring us to be ever-ready to adapt, restart, and restrategize as markets evolve. However, our interpretations and psychological responses to these events can only be effective if we direct our energy into purposeful actions, enabling us to handle the mental gymnastics required to understand these trends. Deep exploration of our brain's intricate neural networks becomes an essential component of this process.

Gleaning from the insights in Chapter 6, we concluded that by carefully modifying the function of our reticular activating system (RAS) in a targeted manner, we can refine our focus. This allows us to filter specific information and solutions and bolster adaptability. To do this, however, we must masterfully negotiate the other facet of the RAS: governing our fight-or-flight response.

Imagine, if you will, the wonders that could be discovered if we could peer directly into the workings of the brain. Moving away from the realm of dreams and into reality, we now have a way to enhance our energy by better understanding our brain. Progress in AI underscores a vital facet of brain function that demands our attention.

The New York Times reported in May 2023 that scientists at the University of Texas at Austin had combined a language decoder AI with fMRI scans—diagnostic tools that monitor blood flow to various brain regions—to effectively read minds. So far the AI has only about a 50 percent accuracy rate, and the person whose mind is being read must be in the MRI machine for it to work, the research shows promise for people unable to speak due to paralysis, strokes, or other medical problems.[4]

Interestingly, the blood flow patterns are significant not only to the production of energy, but also in our initial reactions to AI. During intense emotional states, blood and oxygen are directed to the amygdala, the part of the brain responsible for the fight-or-flight response, diverting resources from the prefrontal cortex. This brain region oversees decision making, problem solving, and predicting the implications of your actions.[5] The RAS, on top of highlighting valuable information, also governs our fight-or-flight response by discarding nonessential information and permitting only information that aligns with our current focus, thus helping to conserve energy[6] In our adaptation journey, conserving energy, regulating cerebral blood flow, and mitigating inflammation are vital mechanisms.

In *Unstoppable*, I compared experiencing burnout to approaching a T-junction. To fulfill your goals, you must turn right, but energy is in short supply. Waning resources send your amygdala into high alert, urging you to turn left toward rest and recovery. As a result, your dwindling energy is channeled toward essential bodily functions and low-effort tasks, sidelining your ambitions. Behavioral patterns such as procrastination are triggered to conserve energy. Far from a sign of weakness, procrastination serves as a barometer, indicating that your body and brain might be operating outside their window of tolerance. You can view this as a biobehavioral response to depletion. Depending on the trigger, the amygdala's reaction can vary greatly in intensity. For instance, in a less intense response, it may induce feelings of slight fatigue or lethargy, prompting you subtly towards rest. On the other hand, in high-intensity reactions, it might interpret energy lows as severe, life-threatening scenarios, thereby enforcing a mandatory rest period through a near-overwhelming desire to sleep. This range of reactions is the amygdala's way to enforce energy conservation and replenishment.

This intersection between ambition and exhaustion creates conflict between your current self and the person you want to become. While the spirit longs for accomplishment, the brain and body are drained and at odds. This tension culminates in cognitive dissonance—a state marked by conflicting beliefs. Individuals may be strongly driven to attain their objectives, but struggle to pursue them due to crippling fatigue. This complex interplay between ambition and biological limitations illustrates the challenge of practicing self-care while adapting to change.

Just as AI uses fMRI scans to predict the workings of the human brain through blood flow, we can predict our actions by tracking the ebb and flow of our energy levels. In this scenario, fostering AI literacy to promote normalization is crucial, but it is equally important to expose ourselves to both positive and negative developments in AI so we can understand and test our psychological reactions to it. Understanding it can help prevent the fight-or-flight reflex from kicking in, maintaining our focus, and better using what energy we do have. The subsequent phase involves a scavenger hunt amidst a city crippled by the storm surge and power outages.

Just awakened at 10am, with the storm finally behind us and no damage to our building, Alex turns to me, asking, "What's next?"

"First, we'll track down an open convenience store to pick up canned goods for our gas stove and some candles, then seek out a Starbucks with Wi-Fi to reschedule our flights," I reply. Descending the five-story brownstone and arriving on Bleecker Street, we miraculously stumble upon an open store, powerless but operating. Using my phone's flashlight to see, I cut through the darkness and the gathering, scanning the shelves in search of food. Predictably, I encounter almost barren shelves; what was left was mostly highly processed foods bursting with sugar and a few cans of baked beans. "Baked beans aren't my friend," I muse to Alex. But it will have to do.

In possession of a plastic bag filled with chips, beans, and cookies, but no candles, we join the many other New Yorkers trying to assess the aftermath and look for an open eatery. After all, we had just walked almost two miles in the cold morning air. Since the power and our cell phone reception had gone out the previous night, we had been unplugged from the world, devoid of any news. Advancing towards Times Square, our eyes catch sight of an apartment building with its outer wall sheared off, making the apartments within resemble a dollhouse. We were operating purely on caffeine, adrenaline, and candy, and I could sense my impending crash. The cumulative stress from the events of the past week was amassing, and my usual drive and optimistic spirit had been supplanted by a sheer instinct for survival. The strain and sleep deprivation were causing neuroinflammation, and my mind felt foggy, reminiscent of the heavy, dense clouds left behind by the remnants of the hurricane. The inflammation was inhibiting my neurons from producing energy, decreasing my brain endurance, and making it increasingly difficult to focus on devising an escape from this city.[7]

Regardless of whether it was 2012 or 2018, to activate my brain as if flicking on a light switch, it was crucial to decrease inflammation and sequentially boost my energy levels. To do so, I needed to reevaluate the data from our survey and seek counsel from an expert in advanced clinical nutrition, who led me to some invaluable discoveries. Little did he know that he helped me write three books over three years.

For nearly 15 years, Dr. Gregory Kelly served as an editor at *Alternative Medicine Review*, which published peer-reviewed research in the fields of complementary and alternative medicine (CAM). He is now senior director of product development at Neurohacker Collective, overseeing the development, pre-commercial research, and scientific support of new and existing dietary formulations. With multiple articles on natural medicine and nutrition, three chapters contributed to the *Textbook of Natural Medicine*, and more than 30 published journal articles, his contributions to the field are immense. But it was his work in the nootropics field at Neurohacker Collective that caught my attention in 2018, when career-ending fatigue led to my speech becoming slurred and my focus scattered.

Nootropics, commonly called "smart drugs," are compounds that can improve brain functions and augment memory, concentration, attentiveness, drive, relaxation, mood, vigilance, resilience to stress, and more.[8] One well-known nootropic, caffeine, is consumed daily by millions. There are two primary categories: synthetic-based nootropics, which usually require a prescription, and naturally derived nootropics, which are developed from natural sources such as plants. These are my preferred option when possible. There are also "nootropic stacks," which refers to a blend of dietary enhancements or nootropic elements designed to work together to attain a specific outcome, like the ones Kelly has developed.

I believe nootropics and other biological qualities like frequency and ultradian rhythms are among the keys that can help us unlock the vast hidden library of knowledge and focus within ourselves—much like the one buried under Bryant Park in New York. And that belief is reasonable: In an interview, Kelly told me, "The brain is a voracious consumer of energy, accounting for approximately 20 percent of our body's daily energy supply. It tends to suffer most significantly when resources are scarce, particularly those involved in building, recycling, and fueling neurotransmitters—crucial components for cognitive tasks and skills the brain does, like focus, motivation, memory, and mood."

What we label as "energy" is, in fact, a molecule named adenosine triphosphate (ATP), which is generated by minuscule cell structures called mitochondria.[9] The primary function of ATP is to store energy and then distribute it to cells throughout the body and brain. Often dubbed the "energy currency" of cells, ATP supplies essential energy for diverse cellular processes, including brain functions.[10] Any imbalance in ATP levels could affect the health of the brain.[11] ATP is also integral in preserving cognitive functions and facilitating several cognitive processes, such as memory, learning, and information processing.[12] Naturally, obtaining sufficient sleep and eating a nutritious diet with ample fatty acids and protein can enhance ATP levels. However, our 2018 survey of 70,000 people shows that additional nutritional support is necessary, given the deteriorating quality of our diets and the increasing demands on our cognitive functions.

Of those who managed to eat a healthy diet 90 percent of the time, 77 percent were peak performers and only 14 percent were underperformers. The peak performers had a significantly lower chance of encountering brain fog, just 12 percent, while the underperformers had a 98 percent chance of experiencing it. Kelly's extensive research on nootropic supplements highlighted one that contributes to ATP level enhancement. Notably, one of these ingredients is a proprietary form of citicoline, which has undergone clinical trials demonstrating its significant impact on improving mental energy, focus, and attention.[13] One study revealed that supplementing with citicoline could amplify mental energy, showing increased ATP energy production in brain cells by as much as 13.6 percent.[14]

A fascinating synergy comes into play when citicoline heightens our energy through increased ATP levels and habitual caffeine consumers try to spark a surge in brain activity. It is crucial to channel this energy effectively to reach the optimal arousal zone for focus. Striking the perfect balance is critical: We require adequate stimulation for razor-sharp focus without being so overstimulated that we lose control of our thoughts. Enter L-theanine, a nootropic Kelly introduced me to a few years ago, which is adept at serving this function. I recommend it to fellow entrepreneurs weekly, whether separately or in a comprehensive nootropic stack. Interestingly, it also adjusts the frequency within our brains, a mission we began in the last chapter, to help us achieve the coveted state of flow.

Consumed unknowingly by millions daily in black and green tea, L-theanine is a well-researched nootropic. As an amino acid, it has been used for millennia to promote brain health, mitigate anxiety, and enhance attention. L-theanine can potentially influence the levels of certain brain chemicals, including serotonin and dopamine, which are crucial for regulating mood, motivation, sleep, and emotional responses. It can also affect cortisol levels, helping us cope more effectively with stress. Its distinctive advantage is that it induces relaxation without sedation. L-theanine manages this feat by boosting alpha brain waves. As indicated in a study titled "200 mg of Zen," taking 200 mg of L-theanine enhanced alpha brain waves within 30 minutes, creating an alert yet relaxed state that was ideal for focus.[15] Yet the real benefit comes when you consume it with moderate, not excessive amounts of caffeine.

It took me years to understand that the origin of my anxiety, depression, and fatigue wasn't psychological, but rather biological. Caffeine, a psychoactive substance, is often promoted as a lifeline to combat fatigue, but its reputation remains contradictory. While caffeine profoundly impacts the central nervous system, influencing brain function, mood, and behavior, it can also amplify negative effects such as anger, irritability, anxiety, and fatigue in people who are particularly sensitive to it—including me.

Before my trip to New York, I never drank coffee. My mood and energy levels remained stable, and I didn't suffer from those midafternoon energy slumps. But when I suffered from jet lag, coffee quickly became a crutch. Yet caffeine is the last thing we suspect when our energy levels drop.

Caffeine can mimic symptoms of anxiety and trigger brain fog by interacting with adenosine levels in the brain. Crucial to the central nervous system, adenosine regulates our sleep-wake cycle. When we're awake, adenosine levels gradually rise, slowly increasing our need for sleep by suppressing the activity of cells in the basal forebrain. Once we fall asleep, the levels decrease.

Caffeine interferes with this process by obstructing the ability of adenosine receptors to receive adenosine, but it doesn't stop adenosine from being produced. Consequently, when the influence of caffeine dissipates, there's a surge of adenosine ready to bind to its receptors, which results in an abrupt "crash" we're all familiar with.

Drinking more caffeine exacerbates this subsequent decline. However, introducing L-theanine into the equation forms a synergistic bond with caffeine that counteracts its undesirable side effects. Research indicates an optimal caffeine-to-L-theanine ratio of 1:2. So for every 200 mg of L-theanine, take approximately 100 mg of caffeine.[16] A practical guideline is to ingest twice as much L-theanine than caffeine.

Through our investigation into caffeine consumption, carried out via our comprehensive survey, we stumbled upon a series of unexpected surprises. Only a tiny fraction (10 percent) of the top 6 percent of peak performers depended on caffeine to get through their day, starkly contrasting with the significant majority (73 percent) of underperformers who relied on it. As a result of their enhanced energy and emotional stability—an often undervalued yet critical element in their success—these top performers recorded a motivation level 81 percent higher than the underperformers. Moreover, their resilience against criticism was elevated, largely due to their mastery of biology and psychology.

This is why so many nootropic stacks, including those developed with the assistance of Kelly at Neurohacker, frequently include L-theanine to offset the negative effects of caffeine. A combination of nutrients is required to thread the needle, reduce neuroinflammation, and achieve focus. Now, when I find myself unfocused, anxious, or depressed, my immediate reaction isn't to assume I lack willpower. Instead, I ask myself, "Am I experiencing inflammation, and what steps can I take to alleviate it?" This is where we land at the intersection between stress, inflammation, focus, and diet.

While the discussion of productivity often excludes nutrition, an emerging field known as nutritional psychology is gradually gaining recognition. Each day, we encounter numerous stressful situations: deadlines, persistent stress, pollution, financial uncertainties, behavior-altering medication side effects, and nutritional deficiencies that can hinder our ability to maintain focus. Nutritional psychology delves into the relationship between food intake, diet, mood, and cognitive function. It underscores the significant interaction between nutrition and mental health. Neuroscience has demonstrated that specific nutrients like omega-3 fatty acids and B vitamins significantly contribute to brain development and the production of neurotransmitters. Should your diet be deficient in these crucial nutrients, it could result in the brain functioning poorly, leading to emotional irregularities and mood swings. Nootropics, along with an anti-inflammatory diet, have the potential to fill the gaps in these nutritional deficiencies while reducing the brain fog that inhibits our focus.

Considering the outcomes experienced by our members, it is frequently observed that psychological symptoms stemming from inflammatory factors and nutritional deficiencies tend to alleviate naturally once these issues have been resolved. Our prevailing health model is, in fact, inverted. Just imagine trying to unravel childhood trauma (or cope with the erratic predictions of AI's future) while your brain is acting erratically and you're struggling to form a coherent sentence. That was me in 2015. But then, thankfully, I had a revelation that would change my life: I wasn't tired because I was depressed; I was depressed because I was tired.

Once we work out the root causes of fatigue and reenergize the brain, the psychological tasks needed to navigate AI's challenges become manageable. But back in New York in 2012, I wasn't aware of the discourse occurring within my brain. I was gulping down coffee as if it was threatened with imminent extinction. Unraveling the various factors responsible for my downward spiral would take years.

Despite the illuminating lights, Times Square looked like it came straight from the apocalypse, with a handful of disoriented people wandering around, still reeling from the previous night's storm. The National Hurricane Center reports that Hurricane Sandy was directly accountable for a minimum of 147 deaths in the United States, Canada, and the Caribbean.[17] Forty-eight of those deaths were in New York and 12 in New Jersey, where it devastated seaside communities.[18]

Upon locating a pharmacy, Alex and I start scouring the racks for candles. I burst into laughter as I grab the lone candle remaining on the shelf. Alex joins in as I exclaim, "It's fucking cinnamon-scented!" He responds, still laughing, "I guess this will keep us warm tonight." Armed with our aromatic candle, we navigate our way to Starbucks—the very place where this adventure began. Like the first time, I stand outside, attempting to connect to their Wi-Fi through the window. But this time, the store is closed. A pattern is starting to emerge. The hurricane is over, but a new storm front is on the way, one that just might break the camel's back.

CHAPTER 8
ADVENTURER'S HANDBOOK
Rule #8: Boost Your Brain Power

Succeeding in our swiftly changing world demands a careful combination of elements such as recovery, the mindful application of nootropics, and an informed understanding of nutritional psychology. Through this balance, you can enrich your focus, maintain poise, and seamlessly adjust to changes.

Harmonizing Rhythms and Boosting Brainpower

1. **Harmonize with your ultradian rhythm:** Familiarize yourself with your natural cycle of productivity and rest, scheduling your most demanding tasks during your periods of peak focus and energy.
2. **Amplify performance with nootropics:** Make nootropics, commonly known as "smart drugs," a vital part of your daily routine. They are crucial to enhancing cognitive functions, sharpening focus, fueling

motivation, and managing stress—all critical if you want to be able to adapt to change quickly and effectively.
3. **Gain clarity through nutritional psychology:** Maintain a daily journal of your energy fluctuations, mood shifts, and dietary habits. Leverage resources such as *The Unstoppable Journal*, available on Amazon, to spot dietary patterns and triggers, like excessive caffeine intake that impacts your motivation and focus.

The path to upgrading your brain and unlocking its full potential isn't about improving overnight, but a carefully orchestrated combination of getting in tune with your body's natural ultradian rhythm, using nootropics for improved cognitive performance, and understanding the essential role of nutritional psychology. Implement this strategy and see how these changes can unleash new levels of adaptability and resilience within you, preparing you for the fast-paced world around you.

RULE #9

Master the Art of Intuition

"**F**UCK IT—WHAT'S THE WORST that could happen?" I whisper to myself. For four relentless days, I had endured the media summit, and nothing had gone my way. My meticulously crafted and rehearsed pitches weren't landing, and, with a ticking clock showing that only a precious 15 minutes remained and the media preparing for their departure, it was now or never.

Casting aside all caution, I approach the representative of one of the world's largest publishing houses, whose strong Brooklyn accent only amplified his daunting persona. For four days, I had wanted to introduce myself to him, but fear of the unknown had held me back. But now, armed with nothing more than my intuition, I begin my pitch.

As my words spill forth in a jumbled mess, he leans back in his chair, expressionless. I wonder if he is taking in my pitch or has fallen into apathy, due to the barrage of pitches he's been subjected to over the past week. I hadn't gotten past my introduction before he begins gathering his folders to leave. "Oh, shit!" I think. I bid him farewell by offering my business card and thank him for his time. As I retreat, I wonder, "I gave it my all, but perhaps that just isn't enough."

With a laugh, I told Alex about my last-ditch effort at the summit. The hurricane made the media summit seem like nothing more than an illusion. But now, as we savored the first real meal in the 48 hours since the storm hit, I had a chance to go over the uncertainties and thoughts that had haunted me during that time.

Despite the cloaking darkness and silence that still had Manhattan in its grip from the Financial District to Times Square, we had stumbled upon a glimmer of warmth—an open, candlelit restaurant standing resilient in the shadows of Greenwich Village. Tonight's dinner was heated not by a kitchen stove, but by a lengthy extension cord connecting a microwave to a car battery. The vehicle, parked halfway on the sidewalk, sat with its hood yawning open, displaying the impromptu lifeline that promised us a warm meal. But after the storm, it was the best reheated pasta I have ever eaten. The crowning touch was the live music, played masterfully on a polished grand piano. In this unexpected encounter, we found a kind of magic—a serene evening in the wake of the storm to debrief. It encapsulated the romanticism often spoken about the Great Depression in this borough.

During that time, Greenwich Village exuded excitement and allure despite the difficult circumstances. Widespread financial constraints couldn't dampen the spirits of its lively youth. As the curtain of the Prohibition era began to fall, it left behind a legacy marked by the rise of homemade bathtub gin and the vibrant blossoming of bohemian culture. With a community nurturing artistic creativity, the Village thrived amidst the turmoil, offering a vital respite from reality.[1]

My longing for escape didn't take long to reach a fever pitch. As Alex prepared to depart at dawn, my travel plans were set back two more days. In the meantime, the city was slowly coming to life. Bars illuminated by soft candlelight started receiving patrons. Renowned Greenwich Village pizzerias flung open their doors once again, their interiors dimly lit by the inviting warmth of their pizza ovens.

New Yorkers, imbued with an unyielding spirit, sprang into action to revive the city and deliver aid to those affected by the storm, reminiscent of their united efforts after 9/11. Once again, they came together in a display of resilience and camaraderie. The hurricane didn't merely bring devastation; it also stimulated a renewed sense of purpose.

Soothed by the live music and granted relief from the sounds of city traffic that had yet to return, due to half the island being without functioning traffic lights, I could finally hear my thoughts, like I had that day in the Rose Main Reading Room. They were no longer drowned out by the city's constant hum or overshadowed by my ambition.

THE WOLF IS AT THE DOOR

Echoing within my mind, the words "let go" made their presence known. The illusion of control I believed I had over my life and this trip vanished, and I was finally present again. After all, what's the worst that could happen? A hurricane hits! Already did!

Throughout my 20s and 30s, I was snared in the illusion of absolute control, persistently aiming to govern every aspect of my existence. I pushed my personal boundaries and tried to keep up with each trend and challenge. But when our power was cut off and I was disconnected from the relentless cycle, my feeling was more profound than just relief. I could finally hear the call of my intuition that I had buried in my busyness.

The persistent nagging of my intuition has also shaped my understanding of AI today, after several months of deep research and reflection, which varied from hopeful to bleak. My main concerns aren't exactly about AI; they're more about yielding my center to disruptive, complex change that is happening at an unprecedented pace. The stewards of this transition—influential figures in the tech industry who don't hold an elected position—are on the brink of causing unanticipated global disruptions. Moreover, the timeline of this shift is not defined, and the world is not paying enough attention to the arising issues and opportunities.

To embrace these changes unconditionally and to have a solely optimistic outlook would mean having to mentally and emotionally wrestle myself into agreement. My secondary fear doesn't stem from the certainty of change but from my ability to relinquish control to the present moment so I can enjoy each day.

In New York, I pursued relevance; when it comes to AI, I crave resolution. I want a definitive answer that I can sink my teeth into. Instead, I find myself in a deadlock. There isn't one. It's like an ambiguous movie ending that leaves you more puzzled as you exit the theater than when you entered.

Having blind trust that AI is a net positive for humanity requires faith in companies that have proved their priorities are profits over people. And they are leveraging one aspect of human nature to replicate and capitalize on: pattern recognition, the technological equivalent of what we call intuition.

Rather than contorting ourselves to conform to a rosy narrative, we should heed our intuition to discern whether we are on the right path. Einstein astutely reflects on this belief, "The intuitive mind is a sacred gift, and the rational mind is a faithful servant. We have created a society that honors the servant and has forgotten the gift."[2] This critique suggests that by favoring logical reasoning and frequently overlooking our intuitive capabilities, we may be sidelining an essential dimension of our human experience and depriving ourselves of the singular insights that intuition can offer.

Human intuition has long been seen as a form of pattern recognition. Our brains have evolved to recognize patterns outside subconscious thought, subtly picking up on hints and gathering information from our peripheral awareness. Understanding that intuition is an intrinsic part of our brain's information processing, rather than a mystical signal from the unknown, doesn't diminish its significance in maintaining a healthy emotional equilibrium.

Psychiatrist Peter C. Whybrow acknowledges intuition's vital role in fine-tuning our minds and interprets intuition as an instinctive self-awareness guided by a preconscious neural network.[3] Preconscious refers to mental processes or information that, although not immediately visible in our conscious awareness, can enter our conscious mind in the future.

This neural network collects information from formerly established patterns and influences fear, hope, focus, ethics, habits, and belief systems. Through practice, like mastering swimming or a new language, it eliminates doubts, freeing cognitive resources for intuitive guidance and attention. As familiar patterns emerge, our intuition guides us onto the optimal path or alerts us to potential missteps.

As information secures its place in our long-term memory, many elements influence our capacity to retrieve that information. Variables such as age of memory, emotional links, applicability, repetition, exposure, and impact are among the primary factors contributing to deep memory retrieval. The irony is AI is monetizing intuition while we still mock those who trust it.

Conversely, the machine equivalent relies on pattern recognition to identify patterns within a massive trove of data and then uses it to make informed decisions or predictions through algorithms. This ability is an integral aspect of contemporary AI systems and is used in a variety of applications including determining what to present in your social media feed, predicting future consumer purchases in e-commerce, optimizing navigation routes in transportation logistics, predicting stock market trends, assisting in diagnosing illnesses in health care, and even analyzing patterns in climate data to make predictions for weather forecasting. Pattern recognition processes incoming data and looks for regularities, which it uses to generate forecasts, categorize information, and refine decision-making mechanisms.

At this moment, our brains are striving to detect patterns in AI's rapid evolution to help us make well-informed and intuitive decisions. However, the abundance of irregularities and inconsistencies hinders our brains from reaching a resolution with a sense of certainty or trust.

Amid the clamor and plethora of user case studies, it becomes challenging to dismantle the preexisting mental patterns formed under the influence of science fiction and past news stories as we close the gap between fiction and reality. Indeed, as we've discovered, certain patterns demand our immediate attention. One such pattern was pointed out to me in a spectacular fashion.

In a spirited discourse, I spoke with Jason Feifer, a notable figure in the editorial world. Formerly the editor for *Men's Health* magazine, he currently helms *Entrepreneur* magazine. Feifer has a novel perspective: He believes the true gift of AI rests in its disruptive power and ability to shake up existing systems.

He used the legal profession as an example—specifically, the current practice of billable hours, saying, "If their work becomes efficient, they can't bill as many hours, and that's what they're worried about... that's a positive thing."

He added, "Why do we still have billable hours? Because nobody was incentivized to make a change because that was the system... But guess what? Now there is, because we are going to break things that are broken, which allows us to build from what matters." Feifer believes this reconstruction of broken systems is "going to be the true gift of AI"—a gift akin to the personal transformation experienced during periods of grief, which we explored in Chapter 6.[4]

Feifer's perspective holds some truth: We have constructed barriers around social and economic frameworks that both sustain and confine us, and we will require extraordinary incentives and momentum to fix them. Pattern recognition allows us to spot the common threads within the problem—and the possibility.

The emerging patterns have been more of a yell than a murmur. For example, I've already started to run some trials incorporating AI into my daily life. My goal was to explore the full extent of what was possible with AI.

By the time this book is released, my editor, full-time marketing manager, and video editor will have been replaced by AI. But first, I will enable them to reskill and adapt to their changing industries. An immigration attorney and, surprisingly, a veterinarian have also been partially supplanted. When my dog, Mitch, experienced distressing digestive issues, I turned to AI for help. By uploading his blood work and running a hypothetical analysis, AI rapidly identified a solution that had eluded three different veterinarians and cost more than $1,000. Within two weeks, Mitch's issues were resolved. It also successfully diagnosed my partner's excruciating stomach pains, a diagnosis later confirmed by a doctor.

See the pattern in these very rudimentary experiments? Traditionally carried out by highly skilled workers, cognitive tasks are already being replaced at a micro level. The advent of AI will downsize large teams as it diminishes the need for a village to run a company. It will empower solo entrepreneurs by arming them with scalable tools that simplify challenging cognitive tasks, alter the wealth structure of the economy, and create a divide where some will forge ahead while others get left behind.

However, this is only a surface-level pattern. A fundamental principle is that people will always be willing to pay for increased convenience, seeking faster, easier, cheaper, and more intuitive ways to get things done. If your business or industry faces various challenges, it is already fertile ground for AI adoption. These can include difficulties attracting and retaining the right talent, rising costs that put pressure on profit margins, and supply chain disruptions. Other obstacles can involve declining productivity, impersonal customer service, wavering customer loyalty, and ballooning marketing expenses. AI adoption can provide valuable solutions and help overcome such challenges in this context.

My intuition, however, was telling me that something even more profound and challenging is at play here. I have grappled with it for months as I desperately attempted to find an optimistic perspective, which leads me to ask: Is the system the broken piece of the puzzle, or are we?

Contemplating the prospect of being replaced by a digital counterpart is distressing. It showcases one of humanity's most standout traits: the knack for avoidance. We will go to great lengths to dodge pain and numb our emotions. As such, could AI represent the double-edged sword that is OxyContin for our generation? Much like the notorious pill, peddled not by Big Pharma but by Big Tech, AI might promise to alleviate our discomforts without addressing the root cause. It could simply swap one existential crisis—becoming overly dependent—for another—being made redundant. This pattern echoes the crisis I experienced while navigating the chaos of Manhattan's streets in a cab ride from hell, where using a convenience to get away from a problem only leads to other issues.

"Slow down!" I yell at the driver. The streetlights are still out, and there isn't a traffic conductor anywhere to be seen. I believe we are going to be T-boned at any second. He erratically pulls over every few blocks and yells at passersby, "Hey, where are you going!?" before quickly speeding off again. (The mayor has issued a mandate for taxis to carry multiple passengers, due to the subway still being underwater.)

Clinging tightly to the handrail, I am en route to yet another hotel. I can count on one hand the number of beds I've slept in over the past 10 days—four, to be precise. In the latest one, I had awakened to find bedbug bites all over my legs. At least this new hotel has electricity, though strangely, only scalding water flows from its taps. "OK, OK, I get it!" I say. "Wherever you go, there you are. I give up already."

I had begun to accept the gravity of my situation. I undertook a two-week national tour in Australia, and just 48 hours after finishing the tour, I left for the Big Apple. It was then that my business partner, responsible for managing the leads I generated, mysteriously vanished. With my business hemorrhaging funds and issues I had attempted to deal with before my trip worsening, I knew I had to take drastic measures, including severing ties with my partner and completely restructuring my business. I was finally ready to accept that my business was broken. It was time to rip off the bandage. The week's events had provided one hell of a distraction, but things were finally starting to settle down. I was too close to the problem because I was the problem! Like many other people, I had failed to trust my intuition by dutifully ignoring it.

Graham Wallas, an English social psychologist, educator, and co-founder of the London School of Economics, proposed a way to address our challenges in 1926, just three years before the Great Depression. This man, with his strong nose, long sideburns, and sleek black suit made a distinct impression on everyone he encountered.

In Wallas' model of the creative process, there are four stages. In the first stage, the focus is on preparation. This includes diving deep into the problem, reading about it extensively in previous works, and making first attempts at solving it, which may initially end in failure. This period may last between weeks to months and concludes when no progress has occurred.

This stage is then succeeded by an incubation period, during which the problem, counterintuitively, appears to be consciously abandoned. It is an unconscious stage of problem solving that fuels intuition. During incubation, an intuitive idea can mature into a real insight. This stage significantly boosts the likelihood of resolving a problem.[5] Remarkably, longer incubation periods paired with minimal cognitive tasks, such as walking or listening to music, often yield more fruitful results.[6] Consider an entrepreneur seeking a pivotal strategy for their startup. After long, fruitless brainstorming sessions, they take a mental breather, shifting focus to simpler tasks such as organizing paperwork or prepping for minor projects. As they rest their conscious mind, their subconscious continues to process the problem. Unexpectedly, the solution to their business challenge suddenly becomes clear during a coffee break or weekend walk in the park.

This period of incubation does not end until a sudden moment of illumination comes, presenting you with one or many solutions to your problem. Illumination can manifest spontaneously during a dream, a casual conversation with friends about seemingly unrelated topics, watching a movie, or after a hurricane.

Following the stages of preparation, incubation, and illumination is the verification stage. In this phase you must confirm the validity of your idea by demonstrating your solution. Illumination can sometimes send you down the wrong path by assuming the validity of insights that cannot be derived from facts. In which case you must start all over again. Just like I had to write this chapter five times. Why?

I hadn't given myself enough time for the incubation stage for my ideas to simmer and coalesce. I've always believed that every nonfiction book should end on a positive note, providing a resolution that's as satisfying as a well-crafted ending in a novel. But the more I researched, the more I validated my concerns, even though I wanted and still desperately want to be wrong. Finally, after weeks of frustration, I checked myself into a hotel for four days to incubate. That's when I finally had my moment of illumination.

If—and that's still an if—AI does swap one existential crisis, like climate change, health care, and the attention economy, for another one, like mass unemployment, it does not take away our purpose; it transforms it. It is within each of these crises that imbue our existence with profound meaning.

Humanity has historically demonstrated the courage to tackle these obstacles, integrating both adverse and positive experiences into our journey. Our psychological growth is directly tethered to the challenges we choose to meet head-on with prior preparation in mind. For example, according to neurologist Lisa M. Shulman, "Grief is a normal protective process. This process is an evolutionary adaptation to promote survival in the face of emotional trauma."[7] We inherently have adaptive mechanisms available to help us, but we frequently lack a tranquil environment that would allow us to tune into them, so we can hear the following steps as clearly as we hear the disruption.

Resilience arises from the challenges we face as we attempt to foster humanity's continued development on a positive trajectory. Conceivably, AI could unearth undiscovered depths of human creativity by prompting us to confront questions we typically suppress and stimulating inquiries into the broader significance of existence. This would drive us to emphasize our innate empathy and hopefulness, while taking steps to address the broken aspects within us.

Whether it's the symbolic act of raising a Christmas tree at Rockefeller Center as a radiant token of hope or enacting a mandate to share cabs to promote a sense of collective responsibility, we are prompted to question the kind of world we are creating. But one question, the biggest one, remains unanswered: If AI is supposed to free us to pursue our creative endeavors, how will those made "free" pay for food and rent? It's the fly in the Metamucil that leaves people mentally constipated when they try to answer it.

Assuming that AI will give rise to new industries and job opportunities also implies that these newly created positions would be beyond AI's capabilities, requiring humans to step in and do what it can't. But if AI maintains its adaptive trajectory, those jobs will undoubtedly be beyond our own capabilities as well.

The simple truth is that not enough time and distance have passed to allow for illumination to occur at a collective conscious level. We have all been too busy avoiding what is broken in ourselves and in society to deeply consider the problems and possibilities. When people challenge weak analogies based on current events, techno-optimists often label them as "Luddites." This tendency of techno-optimists to name-call serves as their coping and defense mechanism, potentially reflecting an unacknowledged stage of grief and an unwillingness to accept their own vulnerability to the impacts of AI.

They have yet to explore its ramifications, as you have. They remain oblivious that by addressing the problems directly, we can identify what is undesirable and then focus our energy on what is truly important. They are entangled in their own contradictions. While AI was conceived to address preexisting problems, the issues that emerge from its implementation will inspire the development of new solutions and the creation of new industries to tackle those problems. But will it be AI or us that tackles them?

One of the primary worries about generative AI is the veil of secrecy underpinning its development and leveraging copyrighted works as training material, thus violating copyright laws. According to an August 2023 article in *The Atlantic*, the copyrighted works of numerous authors, including notables like Stephen King, Zadie Smith, and Michael Pollan, are being vacuumed up and used by AI, although OpenAI disputes this.[8] Ironically, AI's greatest vulnerability in the end could be AI itself, and not just in an ethical or legal sense.

MASTER THE ART OF INTUITION

A 2023 study by Rice and Stanford University researchers discovered that training AI models on AI-generated content degrades the quality of their outputs. Repeatedly feeding generative AI models—which include both large language models and image generators—with AI-derived content creates a self-consuming cycle that the researchers from the study describe as driving the model "MAD."[9] In other words, without "fresh real data" from original human works to feed the wolf, its output will suffer. Just as we become dependent on it, it is dependent on us—as is the tech industry if copyright lawsuits thwart their endeavors.

The prompts us to question - if our writing, online activities, dialogues, workplace activities, blog entries, and video contents are used to train AI, should there not be some form of compensation provided in return? This is where we come to what is, but not necessarily should be, a politically charged solution. We must now set aside binary thinking and explore the potential within the problem.

The idea of universal basic income (UBI) has garnered new attention from AI titans. They consider it a potential strategy to mitigate the economic repercussions of AI automation. UBI is a 500-year-old policy that is both simple and radical in its design. In this concept, there would be an increase in taxes on major AI entities, collecting the revenue, and then disbursing it as small, perpetual payments to everyone without conditions. Sam Altman, CEO of OpenAI, has a controversial plan to deliver it.

In 2023, Altman created a distinctive cryptocurrency known as Worldcoin, incorporating features akin to a "digital passport." Issued after an in-person iris scan, this passport effectively distinguishes genuine humans from AI bots. At its inception, Worldcoin had already amassed 2 million users from its beta phase and expanded token operations across 35 cities in 20 countries. The blockchain stores these World IDs, maintaining privacy while resisting control by any single entity. Altman hopes that Worldcoin will play a significant role in reshaping the economy influenced by generative AI and facilitate the implementation of UBI by curtailing fraud. He believes it could also help address income inequality.[10]

Such a concept could be labeled a handout in our current polarized political atmosphere. But if our data has contributed to AI companies amassing billions of dollars, is it truly a handout? Or is it a royalty or licensing fee, much like the ones authors or musicians receive when their work is used?

According to Drexel University's Center for Hunger-Free Communities, concern over UBI discouraging work is unfounded. UBI has been implemented internationally, and it didn't make people lazy or erase their purpose; it enhanced it. UBI has even been tried on a small scale in the United States.

Established in 1982, Alaska's Permanent Fund is a sizable nest egg valued at $75.7 billion as of July 2023. The state channels 25 percent of its mineral royalties—profits generated from mining operations, oil fields, and gas reserves—into this fund each year and strategically invests it in various assets. The interest from these investments is then disbursed to every state resident, including children, each September. The payment in 2022 was $3,284, one of the largest payments in history. (So, a family of four received $13,136.)

Research has shown that the Alaska Permanent Fund's payments have not adversely affected employment, and increased part-time work by 17 percent.[11] Additionally, the state experienced a 15 percent rise in entrepreneurial endeavors, which fostered self-sufficiency and generated new employment opportunities.[12] The fund has its challenges, of course. After decades of successful operation, a disruption occurred in 2015 due to a sharp decline in oil prices, leading to a state budget shortage.[13]

Former Governor Jay Hammond, the architect of the fund, designed the dividend system to guarantee that Alaska's non-renewable resources would yield a perpetual return for the state. Implementing UBI to realize a continuous return on renewable human resources is a discussion that will intensify over the coming years, particularly if new jobs do not emerge to replace the ones lost to AI.[14] Instead of mining oil, we are mining intellect.

AI indeed poses a challenge as well as a sweeping opportunity, and we must prepare ourselves for multiple scenarios. Our intuition can guide us if we can take long enough to incubate on the questions we need to ask. If only we could escape the noise of the city and sit quietly long enough for the answers to reveal themselves.

At last, I find myself at LaGuardia Airport, anxiously awaiting my flight to LA and then to Australia. My heart pounds as I watch a long list of flights being canceled, one after another. Hurricane Sandy has already grounded thousands of flights, and now an exceptionally early nor'easter is threatening to dump heavy snow on the regions already ravaged most severely by Sandy. I laugh inwardly, musing, "Here we go again!?" Once again, I surrender my grip on control, and just wait. It is emancipating.

The prospect of the marathon 23-hour flight back home begins to stir my excitement. I plan to bring out my laptop and delve into all possible scenarios for my business, reflect on the trends I have noticed, and use my imagination to create a new strategy. In the chaos surrounding me, I start to ponder an important question—what single goal, if achieved, will change everything?

Then I hear the announcement for my connecting flight: "Now boarding: LaGuardia to Los Angeles." As I step on the plane, another announcement follows, "All remaining flights have been grounded due to a severe weather system moving in."

CHAPTER 9

ADVENTURER'S HANDBOOK

Rule #9: Master the Art of Intuition

Now that you are familiar with the four-stage creative process proposed by Graham Wallas, let's examine its connections with intuitive prowess and pattern recognition. Intuition and pattern recognition are not mere tools for prediction; they are robust skills for observation and making connections. Strengthening these skills can offer you remarkable foresight and a vibrant world view, helping you predict and navigate future possibilities.

Strengthening Prediction Skills with Intuition and Pattern Recognition

1. **Preparation**: Strive to understand the problem by fully investing your efforts into research and exploration. Develop a solid foundation in the subject matter, attempt various solutions, and analyze the problem from multiple perspectives. Equip yourself with ample knowledge so you can make informed decisions and observations.
2. **Incubation**: Following the preparation stage is the incubation stage, where you consciously set aside the problem. This period does not end until there is a revelation. This step is vital for fostering intuition. It is during this incubation period that you engage in the following series of actions:
    - **Elevated mindfulness**: This involves practicing mindfulness exercises, meditation, or simple awareness activities to help enhance subconscious sensitivity to often overlooked cues and formats.

- **Solitude and silence**: By spending quiet, undistracted time alone, you foster an environment that allows intuition to surface. Periods of silence often bring clarity, letting patterns form and intuitive insights to be recognized.
- **Intuitive exercises**: Engage in activities to foster intuition, like "mindful walking," where you engage your senses by being conscious of your environment or using mediums like tarot cards to encourage introspection and pattern recognition.

3. **Illumination**: This is the moment of sudden insight and clarity, where intuitive hunches and recognized patterns come together to form an idea or solution. These "Aha!" moments can emerge unexpectedly, in dreams, during casual conversations, or while you are performing routine activities. Cultivating your intuitive abilities will allow you to perceive connections that might otherwise go unnoticed.
4. **Verification**: In the verification stage, you validate your intuitive ideas and patterns by rigorously testing and demonstrating the solution(s) you found in the illumination stage.
   - Remain adaptable, as intuitive insights may not always be firmly grounded in facts.
   - If that turns out to be the case, retrace your steps through the stages, revising your approach and refining your understanding of the problem.

Incorporating Wallas' four-step approach into your creative process will empower your intuition, sharpen your pattern recognition skills, and prepare you to effectively navigate and predict future possibilities. This holistic approach will help you confidently adapt and flourish in an ever-changing world.

RULE #10

Make Tough Decisions, Fast

AS THE ELEVATOR SWIFTLY SOARS skyward through fifteen floors, my breath maintains a steady, deep rhythm. Anticipation builds within my chest. My business partner, accompanied by a mediator they'd known for years, is there to discuss terms for me to regain the 50 percent company stake I had once carelessly given up to them. I had been sold on the hope of scaling the business, with my partner taking over the marketing and administration, freeing me to become the face of the brand and hone my craft. But all their promises amounted to nothing more than hot air, dissipating into the reality of harsh business negotiations. I had kept my end of the bargain, speaking at 60-plus events a year, running three-day workshop intensives, and going on a national tour before I had departed for New York, but they hadn't kept theirs.

I have just returned from Manhattan a week ago, barely managing to avoid a snowstorm that paralyzed LaGuardia shortly after my flight had taken off. At home, in something of a trance, I try to sleep and watch the news coverage of the hurricane and the snowstorm that incapacitated New York and caused billions in damage. It feels surreal. I can't shake off the residual overwhelm.

The $15,000 I spent on the trip was heavy on my mind. After compulsively refreshing my email for the hundredth time, the complete absence of replies from all the media contacts I had met was glaring; I have failed to book a single media appearance, my whole purpose for attending. Were all the preparation, stress, money, struggle, and sleepless nights in vain? Was my moon shot worth it?

This is supposed to be a turning point when my whole life pivots, and yet it seems as if I have gone back to the starting line. But the challenges confronting me now have sharper teeth. My business partner is demanding a $50,000 payout, which is far beyond my means. The balance of my livelihood teeters precariously. The urgent need to find a solution becomes a desperate race against time.

Life in New York, with lessons of grit and survival that surpassed the rigor of any esteemed Ivy League, ingrains a vital truth: to realize cherished dreams, one must be prepared to brave not just a storm, but a hurricane of challenges that hit all at once with astounding force. High-stakes decisions must be made.

As the elevator door's part, I step out, ready for the battle to reclaim my business built on my toil and tears. In the distance, behind glass doors, my business partner and the mediator sit at a round table. Their evident strategy session before my arrival is a clear attempt to acquire an advantage, but I have anticipated their pregame plan ahead of time. After exchanging platitudes, we sit down.

The mediator outlines how the meeting will unfold: We will each get an opportunity to present our preferred solution before hopefully coming to an agreement. In a strategic move, I plan to reveal my cards last hidden under the veneer of politeness.

My partner persuasively presents their argument for the $50,000 payout, even claiming (falsely) that my trip to New York has been a vacation paid for by the company. Once they finish, the floor is mine. From a folder, I extract a piece of paper. It is an estimate from a marketing company listing all the tasks that have fallen by the wayside over the past several months due to my partner's negligence, resulting in a dramatic revenue drop. The total losses? $53,495.

The mediator's jaw falls as the weight of my words sink in while my partner's lips become a thin line, barely concealing their ire. It seems as if they might lunge at me at any moment. The mediator then asks me to leave the room to give them some privacy, and a few minutes later, he invites me back in, still in disbelief.

When I walk in, my partner asks me a single, terse question: "Ben, what do you want?" In the days before the meeting, I had done my research and come up with three clear options, though I only voice two: "You can complete the unfinished work, valued at $53,495. Alternatively, I can repurchase my shares for $1 each, and we can call it a day."

The third option, which I didn't mention, is this: Under Australian law, in a 50/50 partnership both parties must share equal responsibility in managing the business. This means that their demand for $50K is questionable, since their neglect caused the losses they're trying to claim. It also implies that they could be liable for damages and open themselves up to legal action for failing in their responsibilities.

Emerging from the meeting with a signed share repurchase agreement and $50 out of pocket, I embrace the fresh start I have been craving and reclaim my power. "I guess the New Yorkers' instinct rubbed off on me after all," I muse to myself as the lift takes me back down. The following day marks the relaunch of my business and a bold new chapter in my life, imbuing me with a rekindled sense of determination.

This renewed drive echoes the values instilled in me by my parents. My father taught me the importance of fortitude, while my mother instilled patience and foresight. Despite confronting numerous adversities, including death, drought, fire, and the looming threat of bankruptcy, they showed remarkable determination. Their perseverance culminated in the expansion of the family farm into a large cattle feedlot accommodating more than 35,000 cattle. What began as a small property more than a century ago bloomed into an expansive 5,000-acre or 2,023-hectare property. For context, the average farm in the U.S. in 2022 was 446 acres.[1]

While the enduring legacy of the Angel family was shown by more than 200 people turning up at Dad's wake—and a local dirt road proudly bearing our name—it's further immortalized in the quiet strength of my father. I only ever saw him exercise his authority once, provoked after ceaseless instigations. However, he opted for strategy over fury, surprising those who underestimated his resolve. His name, Steele Angel, was a fitting testament to his unyielding fortitude.

Upon gazing at the promise, penned at the bedside of grief, which now graces my home office wall, I feel a strong resolve to reinvent myself. Each morning, I dedicate an hour to learning everything I can about the latest trends in digital marketing. I then make a rule that I must implement at least one strategy per day so I can absorb the lesson and understand its implications. As I am undergoing this process, a young woman named Amy Porterfield is on her own journey, crafting content on using social media to help market your business—a concept, like today's AI, that we are all struggling to understand. Unknown to me at the time, she played a crucial role in my transformation. Like me, she went through a stage where she had to negotiate her own future. And 10 years later, I finally got the opportunity to speak with her.

A self-described ex-corporate girl who transformed herself into an online marketing expert, Porterfield now sits at the helm of her multimillion-dollar enterprise. She underwent an ordeal similar to mine, but with the stakes raised into the millions. Currently she is the author of the *New York Times* bestseller *Two Weeks Notice* and endured a nearly disastrous year that most people don't know about. Like me, she relinquished a 50 percent stake in her company to a partner, which resulted in her bearing the brunt of the work and clocking more than 60 hours a week. "It literally derailed me for a full year, thinking I'd lose everything," she told me. "It was very, very scary. It was my livelihood. I felt a lot of shame. I chose to bring someone into my business with very little legal contracts, give them 50 percent, and kind of let them run the show. That wasn't his fault; that was mine. I let him call the shots."[2] You might be thinking, wasn't it unwise for Ben and Amy to relinquish half of their respective companies? They both chose blind faith over vigilance. And you would be right! Yet aren't you doing the same thing?

Are you prepared to pass the reins of your future to big tech without demanding strong regulation, due diligence, and complete transparency? Surely you wouldn't want to risk your future without a seat at the negotiating table. These enormous human intelligence, intuition, and creativity trade-offs demand careful negotiation, not blind faith.

AI introduces distinctive challenges to our values and beliefs. We strive for convenience, but not at the expense of our businesses or jobs. We invite cures for diseases but fear a new pandemic. We crave authentic connection with our loved ones, but recoil at the thought of an AI impostor. We yearn to have faith, but dread being manipulated into unlawful acts. We crave quick and precise answers, yet we fear that in seeking such solutions, we might inadvertently harm the industries in which our children are currently honing their skills. AI challenges our values in many deep and complex ways, especially as the 10th threat comes into focus: weaponization. It manifests in three distinct ways. The first is so frightening that sharing it feels almost wrong, but there's a good reason for doing so.

Those threats are "slaughterbots," or, more formally, "lethal autonomous weapons systems." They are weapons systems that use AI to identify, target, and kill people without a human operator.[3] A 2021 UN report suspected the use of LAWS in Libya, and since then these autonomous systems have raised alarm bells worldwide. UN Secretary-General António Guterres called these systems "politically unacceptable, morally repugnant, and should be prohibited by international law."[4] This chilling image of "slaughterbots" on the battlefield brings us closer to home, reflecting on the evolving conversation around using lethal autonomous systems within our own borders. In November 2022, the San Francisco Board of Supervisors debated whether law enforcement robots should be allowed to use lethal force. The discussion came after new California legislation passed, requiring transparency about local law enforcement's use of military-grade equipment. The draft policy initially prohibited the use, but later changed it to allow robots to use lethal force under specific circumstances.[5] The change sparked instant controversy, and many robotics companies issued a public statement that general-purpose robots should not bear weapons, warning against their potential misuse.[6]

The second form of weaponization could trigger a catastrophic market collapse of unprecedented magnitude. AI-powered "black box" trading algorithms pose a significant risk to the stock market. If something went wrong and they all sold the same stock simultaneously, the market crash could surpass the devastation of the Great Depression. SEC chairman Gary Gensler, who has scrutinized AIs' possible consequences for years, warned in 2023 that AI models may favor corporate interests over investors, causing them to offer misleading financial advice. The SEC has therefore proposed a rule requiring AI models to eliminate conflicts of interest from their training data.[7] This scenario connects to another grave concern: The third method of weaponization could unleash the next large-scale pandemic.

In a class experiment at MIT, undergraduates without a background in biology easily obtained detailed recommendations for how to make bioweapons, such as smallpox and the avian flu, from popular chatbots such as ChatGPT. Recognizing the potential for an AI-generated pandemic, responsible figures in the biotech industry are taking this threat seriously. Since its initial launch, ChatGPT has implemented measures to make it difficult for users to acquire blueprints for manufacturing fatal viruses. Ginkgo Bioworks, a prominent biotech firm, has joined forces with U.S. intelligence agencies to create software capable of identifying engineered DNA on a large scale, thus giving investigators the tools they need to trace a manufactured germ. This collaboration shows how cutting-edge technology can safeguard against other, deadlier uses of technology.[8]

Big tech currently stands at the convergence between biology, psychology, and technology, tasked to safeguard our collective future—a challenge with far greater gravity than the decisions Amy and I faced. The ripple effects of their choices run deep, shaping the future and, at times, calling for betting with everything we have. We all play a vital role in the decisions that impact our lives and those around us. I had to make a decision like that 20 years ago; it still haunts me to this day.

THE WOLF IS AT THE DOOR

In the next room, my father lies with a feeding tube as his only lifeline, his voice silenced by his condition. Unknown to me, my mother has summoned my grandmother—my father's mother—and my brother for a conversation that will have monumental consequences. "The doctors don't expect him to survive beyond three to six months. We are facing the difficult decision of whether to withdraw life support," she confesses, her voice trembling with the agony of impending loss. With tears streaming down her face, my grandma replies, "I've already buried my husband and one son. I don't want to bury another."

Upon hearing her words, emotion overwhelmed us all. I hug her before uttering the most challenging words I have had to express in my 23 years of life. "Grandma," I began with a tremor in my voice, "I know this is painful to hear. But I remember when I was a child, Mom and Dad sat us down at the farm table and asked us frankly what we would want if the prognosis were grim: to stay on life support or to be taken off?"

My mom's gaze met mine as I pressed on. "Dad said he didn't want his final days plagued by suffering, or for us to endure the pain of witnessing it." As the realization washed over Grandma, she burst into tears, pulling me into her arms and whispering, "You're right." Mom and my brother nodded in agreement; they too recalled that conversation, which had left an indelible mark.

Weeks after Dad's passing, Mom came to me and tearfully expressed her gratitude for speaking up in that moment. She admitted her difficulty in voicing Dad's wish—faced with the death of the man she loved, her courage had wavered. In his silence, I became his voice. I didn't want him to die, but I also didn't want him to suffer. I had to advocate to give him peace at the expense of my own. It is an impossible choice to make, and sometimes it is even harder to live with. Yet an impossible choice is what advocates of unregulated AI are demanding of us: adapt or die, but then die or lose anyway.

We are frequently encouraged to "embrace AI," but if we do, we may be embracing the very technology that replaces our jobs and our children's future careers. We hear assurances that "every revolution has brought about new jobs," though seldom is there a concrete elaboration concerning the specific jobs that await or which industries will remain unscathed. We often hear that "regulation will stifle innovation," while its potential to promote innovation is largely unexplored. "It's not like the Terminator," they say, but as we have seen, it's closer than we would like. Finally, the most pervasive words of reassurance are "Don't be afraid"—a phrase that fails to encapsulate the complexities and challenges of AI, with its fingers now in every facet of our lives. It is modern-day gaslighting, and it is why our reaction to AI closely resembles the grieving process, due to the impossible choices we are presented with. Equally, we must shoulder the weight of our inaction when we do not have a seat at the table to negotiate our future.

Thankfully, momentum is shifting in humanity's favor, even if that change often gets drowned out by the hopelessness that has been permeating our lives. This shift has been mirrored in the discourse around AI since I began working on this book, particularly as of August 24, 2023. In a defining moment, the UK government announced that Bletchley Park—the renowned site of the World War II code breakers, led by Alan Turing, who cracked Germany's Enigma code in 1941—would host the world's first artificial intelligence safety summit in November 2023.[9]

Turing, often regarded as the father of AI, continues to shape and inspire crucial conversations on ethical AI development and help evaluate risks associated with groundbreaking advances. With judicious regulations and robust support for innovative applications of AI to tackle challenges in science, health, and the economy, we can revolutionize the world in beneficial ways.

Yet there is a common argument made against regulation, and one that is fundamentally flawed, called the Nirvana fallacy. Economist Harold Demsetz coined the term in 1969.[10] The Nirvana fallacy occurs when actual solutions are measured against idealized, unattainable ones, and it's concluded that a solution isn't worth pursuing if it isn't perfect.

One example is the assertion, "Seat belts are pointless, since people who wear them can still die in car crashes." This argument overlooks the fact that fewer people die in car crashes if they are wearing their seat belts, and that incremental progress is a valid form of improvement.

The Nirvana fallacy presents a deceptive either-or choice, with one option appearing better, but fundamentally unattainable. A person can use this fallacy to challenge any opposing idea because of its inherent flaws. Consider the statement, "Establishing regulations on technological innovations will obstruct progress, so we should reject all such regulations." This mistakenly assumes any regulation will hinder innovation, neglecting to point out that it can also responsibly steer technological growth and prevent possible misuse. It also overlooks that leading innovators and creators in the AI industry are advocating for establishing regulations.

Opposing any regulation because it might obstruct some progress is unrealistic. We have established standards and regulations in our financial markets, and they have not obliterated financial innovation; rather, they guide it within permissible boundaries. In the same way, judicious regulation in technology can encourage ethical and sustainable progress.

In a March 2023 interview with *Forbes* magazine, global consultancy firm PwC said that regulations could foster innovation while inspiring the emergence of inventive business models and consumer benefits. They highlighted the example of tech giants IBM and Microsoft, which have both had to face issues with government regulation. PwC said, "Both responded in ways that benefited the industry and consumers, allowing legacy companies to thrive while making room for new players."[11]

The regulatory actions against Microsoft made room for other groundbreaking pioneers such as Apple, Amazon, Google, and Facebook. Apart from encouraging competition and development, regulation can also help legitimize AI in new sectors, with the potential to revolutionize outcomes.

In this discussion, we must also confront our own fallacies. For instance, consider the belief "AI represents an existential threat to humanity because it could spark the next war, crash the markets, or unleash a virus, and therefore we should ban it." This viewpoint argues that protective measures cannot be implemented swiftly enough to prevent the worst-case scenarios. However, contrary to this belief, the European Commission has set forth a new global AI framework. Unveiled in September 2023, this framework is the world's first comprehensive, pro-innovation AI regulation. It rests on three key pillars: guardrails for the technology, governance to oversee its use, and a strong focus on guiding innovation. The European Commission called for mitigating the risk of extinction from AI to be a global priority, demonstrating that proactive measures can be taken.[12]

It's true that some fear the timing of its enforceability may not be fast enough to keep pace with AI advancements. However, this underestimates the dynamic nature of the regulatory adaptations that could happen concurrently with the enforcement. As for global participation, while complete adherence by all nations remains an idyllic scenario, the presence of such a framework sets a much-needed regulatory precedent, encouraging global dialogue and cooperation on this pertinent issue. Lastly, the risk of a rogue AI slipping through, while plausible, is significantly reduced by comprehensive governance and continual vigilance - pillars on which the European Commission's framework rests. While risks exist, there are effective proactive measures that can be taken to mitigate them.

While any solution will be less than perfect, that does not mean we should not look for them, especially when we are facing threats like 1) manipulation, 2) misinformation, 3) AGI/sentience, 4) misuse, 5) automation, 6) information warfare, 7) an attention recession, 8) hypervigilance, 9) irrelevance, and 10) weaponization. We have challenging decisions to make as we navigate our future, both on a personal and societal scale. The question is, do we have any influence over how this technology is used and what it can be trained on?

THE WOLF IS AT THE DOOR

It may be in the legal and political domain of privacy rights where we can best find our equilibrium and gain an advantage in our negotiations to develop AI that benefits everyone. Companies are hastily seeking the right to use people's data for AI training. This data, if mishandled, could be leveraged to manipulate political viewpoints, influence AI policies, and incite division. Diverse companies, from Twitter (now X) and Instacart to Microsoft, Meta, and Zoom, swiftly revised their privacy policies in 2023 to allow using information and content from users and customers as data for training their AI models. Zoom, which experienced a public outcry over the notion of user video calls being ingested by an LLM for training, is the sole company that has since amended its usage policy, which now explicitly states that user videos will not be used in such a manner. The backlash that Zoom faced hasn't deterred other companies from deciding that their platforms should serve as training environments for AI.[13] This should make us mindful of what information we share, especially personal information, and how it may be used against us.

If the pregame plan is to call anyone who disagrees with regulation a Luddite, then the counterplan is to call for nuanced strategy, not fury. We must use our imagination to find solutions; work to understand the dynamic nature of AI; promote ethical practices, policies, and social responsibility; call for transparency; actively participate in discussions at work and home; and encourage lifelong learning and adaptability. Because sometimes, even when the stakes are high and you believe all has been lost, you win.

As I sit engrossed in a new project in the comfort of my home office back in Australia, reveling in the feeling of having won my company back from my former business partner, an incoming call from an unknown number interrupts my concentration. "Ben speaking," I respond, my voice laced with an upbeat casualness, curious about the unfamiliar number. "Sorry, could you please repeat where you're calling from?"

Suddenly it hits me—New York!

"You met our colleague in Manhattan," she replies. "He told us he loved you and to get you in for a meeting ASAP! We want to discuss working with you on a new book." Striving to contain my mounting excitement and not wanting to appear too eager, I say, "Sure, let me check my calendar." Feigning importance to avoid appearing too eager. Clearing my throat, I reply, "It looks like I'm free on Tuesday," in a slightly higher tone than I desire.

One year after getting my business back, I released my first book through a worldwide publisher, launched my online educational programs, and scaled to millions in sales and tens of thousands of members. I sold my beloved Port Melbourne apartment a street back from the beach because I had become too complacent and set off for the U.S. to write *Unstoppable*, which sold more than 80,000 copies. This book you hold in your hands is my seventh one. Meanwhile, Amy catapulted her sales from $5 million to a staggering $16 million, while her podcast audience exploded to more than 55 million downloads. How? Well, that answer leads us to the final stage of the grieving process, which can only be achieved through a technique I have hidden within the pages of this book. It's a strategic omission - as we tiptoe toward it, let's remember this exploration isn't just about AI—it's about you! To understand our future direction, let's retrace our steps. The time has come to get out of this cab ride from hell.

Together, we've dealt with shock and disbelief, contended with anger caused by the disruption of our usual rhythm, and faced the threat that technology may eclipse human labor and affection. We've encountered profound sadness and helplessness and grappled with guilt over contributing to our digital counterparts. We've negotiated these upheavals while reevaluating ourselves and ultimately uncovered hope. But acceptance is a personal journey; not every revelation must be accepted.

Hope isn't simply bestowed on us; it's cultivated through patience and experience. The bridge to hope is built by imagination, complementing the skills we've acquired, like adaptation, AI literacy, meditation, visualization, focus, and more.

Imagination enables hope by painting a captivating vision of future possibilities. This process demands the courage to confront our deepest fears and the grace to balance a spectrum of emotions, from fear and confusion to excitement and optimism.

You may wonder, "Ben, why navigate through all these potential AI pitfalls instead of highlighting just positives?" My approach is primarily layered to accustom you gently to the imminent changes and help establish your tolerance, prepare you for future shocks, and arm you with mitigation tools ahead of the curve. Additionally, it forges a pathway for critical thinking and open dialogue, channeling fear toward sharpening your focus when courage is needed.

But the most important reason is to present it in a manner that sparks your imagination by taking you on an unexpected journey. Bolstering our imaginative abilities requires reflecting on personal experiences, speculating on the future, and visualizing new scenarios. These exercises kindle our brain's imagination network, invoked when our minds wander or daydream. This mental canvas is crucial for creating vivid alternative paths that diverge from our present.

This proactive engagement with imagination fosters inventive problem solving and the discovery of unconventional solutions. It weaves a panorama of possible consequences, breaking down barriers and unveiling fresh answers.

By outlining and articulating your fears as clear as daylight, you can recognize the hazards of inaction, and with a deep understanding of AI, you can effectively blueprint your future. Preparing for undesirable outcomes, such as AI disrupting your job or profession, positions AI not merely as a threat, but as a transformative tool for your career. By avoiding false hope, you can take control of possible avenues to success that could otherwise remain elusive, even if the reality isn't quite what you'd hoped for. This book isn't just a warning; it's your game plan.

In a typical nonfiction book, I would eagerly present a straightforward argument and conclusion and assure you that everything will turn out just fine. This time, I cannot do that—not

MAKE TOUGH DECISIONS, FAST

because I don't want to, but because it's not feasible. What I can say, though, is that none of your challenges can be addressed if you're captivated by the dazzling distractions around you without a clear understanding of who you are and what your goals are. Without focus, you'll be too distracted to prepare; without energy, you'll lose your momentum; without vulnerability, you will become the most vulnerable; without flexibility, you'll snap; and without introspection, you'll never have the courage to do what is right when the moment calls you, even if it means losing.

We have now reached a critical juncture. After delving deep into the challenges that lie ahead and the preparations necessary to face them, it's time to revisit two key questions I asked at the beginning of this book:

When you meet the wolf, will he be friend or foe? Will you tame the beast or ignore his existence at your own peril?

Before you answer these questions, answer this one, do you hear the howling at your door? I do—but it's not the wolf. It's the wind gusts from Hurricane Idalia's distant bands, a late August storm currently on the 30th day of the month in 2023, swaying the palm trees seven stories below. I am sitting upright in bed in a hotel far away from the ocean that is currently flooding our street back home in Tampa, Florida. My trusted companion Mitch, with his honeyed, warm caramel fur, snuggles against my arm, beseeching me with his eyes to turn off the TV. With a contented smile, I acquiesce and ease back onto the pillows.

"Mitch," I begin, gazing at my faithful friend, "I recently realized something. What I pursued in New York wasn't relevance after all, but adaptability. I also believe I've found a middle ground with AI. To be honest, I didn't think I would. I'm a Luddite, damn it! And proud of it."

I don't oppose new technology; I will continue to embrace it. I oppose its misuse, just like the original Luddites did over 200 years ago. I will also continue to be the farmer, dutifully preparing his field of knowledge, applying his creativity, adaptability, and, most importantly, his fortitude, striving to achieve an ideal balance in integrating technology and its positive impact on society. And, if need be, I will stand up to protect that crop of knowledge. But most of all, I will master focus and practice patience because this story is only just getting started and will require ample supplies of each.

As long as I remain hopeful and prepare for the future like I would for a looming hurricane—staying informed about its path and taking all necessary precautions—I believe we can weather this storm and assist those caught in its cross hairs. Because if my parents taught me anything, it is that hope reigns supreme.

As Mitch's little pink tongue lapped at his paw idly, I say aloud, "Mitch, do you think the readers have figured out who the wolf is?" He offers no reply, simply stretching his paws and nestling in. I, too, find myself drawn into the realm of dreams. Yet something peculiar happens, they aren't my dreams—they are yours, dear reader. And at last, in this pulsating metropolis that never sleeps, you will come face-to-face with the wolf.

CHAPTER 10

ADVENTURER'S HANDBOOK

Rule #10: Make Tough Decisions, Fast

Before I reveal who the real wolf is, we should acknowledge the vital role of decision making, particularly as we venture further into the AI age. The ability to make tough, even disagreeable decisions has become a distinguishing factor in an increasingly volatile landscape. With this rapid transformation, the power to make hard choices becomes a necessary quality for individuals and businesses, whether it is about a strategic career move or a vital business decision. Having the courage to face difficult choices head-on has become an indispensable asset, as has looking beyond the current moment when making those choices.

Choosing Battle Lines: Making Moves in the Face of AI Domination

In the AI era, various decision-making categories present unique challenges for individuals and businesses. These include:

- **Preemptive career moves**: Deciding when to change career paths, particularly toward AI-integrated roles
- **Reskilling**: Determining if, when, and how to acquire new skills to stay relevant in an AI-dominated work environment
- **Staff layoffs**: Making hard decisions about workforce restructuring or layoffs due to AI adoption
- **AI adoption**: Deciding to implement AI technologies, knowing it may lead to job losses
- **Worker rights**: Ensuring employees maintain their rights, fair treatment, and ethical considerations when integrating AI in the workplace
- **Examining learning options**: Students must assess whether what they are currently studying may become redundant in the not-too-distant future.
- **Voicing concerns:** The need to protect your intellectual property could compel you to raise your voice, such as when the Writers Guild of America went on strike to thwart AI's potential to suppress salaries.

Sailing Uncharted Waters: Techniques to Tackle Complex Choices

To navigate this new terrain, a four-step process can significantly enhance the effectiveness of these choices:

1. **Situation assessment:** Thoroughly study the decision at hand. Analyze its impacts, potential outcomes, and possible alternatives.
   - What are the potential repercussions of this decision?
   - Is there a preferred choice if multiple options are present?
   - Are there any unanticipated risks?
2. **Deliberation period:** Allow yourself some time and space to consider the decision, akin to the incubation stage. It could involve discussions with stakeholders, further research, or a quiet period of reflection.
   - What insights emerged from the discussions or during the reflection period?
3. **Moment of insight:** Acknowledge that "Eureka" moments can occur during decision making, presenting unforeseen solutions or options.
   - Has a new solution or option emerged that was not initially considered?
4. **Adaptation:** Evaluate your decision, considering every potential outcome. Be ready to iterate, revise, and adapt throughout the process.
   - Does it hold up against the realities of the situation?
   - Have any blind spots been overlooked during the analysis?

By following this rigorous process, tough decisions can be approached systematically and mindfully, ensuring that every potential outcome has been considered and the decision made is both tough and effective.

Decoding the Final Threat of Artificial Intelligence

Of the many perils that AI technology poses to us, we have examined manipulation, misinformation, AGI/sentience, misuse, automation, information warfare, attention recession, hypervigilance, and the looming threat of irrelevance. However, we must confront one final, formidable challenge: weaponization. This omnipresent danger completes our sobering roundup of AI's top 10 threats.

10. **Weaponization:** The potential misuse of AI technology is a grave concern in several areas. Weaponizing robots could unleash an unprecedented scale of destruction that surpasses traditional warfare. AI-induced disruptions of the stock market could spark a financial meltdown or incite severe social unrest, possibly catalyzing civic confrontations and conflicts. AI could also wreak worldwide havoc by helping malignant actors gain access to bioweapons like smallpox or avian flu, leading to the next pandemic. The proliferation of weaponized AI is a sobering prospect that calls for stringent regulation and ethical guidelines to prevent escalation.

Temperature Test

Perform a self-assessment now, acknowledging that your views may shift as technology progresses. Identify where you stand with AI: are you a skeptic, an optimist, a realist, or a doomsayer? Or perhaps you find yourself vacillating between these perspectives? Alternatively, you might identify as a modern-day Luddite, one who does not oppose technology outright, but its misuse.

10 Rules to Survive and Thrive in an AI-Driven World

Life in an AI-controlled world can feel like navigating a maze. The transition may be destabilizing and filled with feelings of loss and discomfort as we leave behind ways of working and living we've known for years. Yet from this grief emerges opportunities to adapt and grow. Within the subtlety of our lived experiences, we can unearth the ability to respond, change, and adapt. A sturdy foundation built on lessons from the past guides us as we walk toward the future.

So, let's draw upon our resilience and review the 10 cardinal rules designed to help you not only navigate but also truly thrive in this rapidly changing, AI-driven world:

1. Expect the unexpected.
2. Avoid cruel optimism.
3. Fuel your focus.
4. Open a window.
5. Accelerate adaptability.
6. Embrace reconstruction.
7. Find your frequency.
8. Boost your brain power.
9. Master the art of intuition.
10. Make tough decisions, fast.

RULE #11

Know Who You Are

As YOU RECLINE in your favorite armchair, a thoughtful expression crosses your face as you consider the ideas shared by Ben. Feeling the weight of the day's stress begin to dissolve, your eyelids give in to the persistent urge to close and your breath grows deeper, signaling the onset of sleep.

With that transition into slumber, your mind flings open the metaphorical doors to its depths, and the familiar comfort of your living room fades, replaced by a surreal landscape that challenges the boundaries of your imagination. It takes a few bewildering moments for you to orient yourself before you dimly recognize your surroundings: You are in the city that never sleeps, New York City.

But you're not standing on the skyscraper-laden streets of Manhattan, but in one of its bustling underground arteries—the Times Square subway station. A rattling hum in the distance signals an approaching train, and you feel an undeniable urge to board it, certain that wherever it is going, it is important.

The grimy platform, dimly illuminated by aging fluorescent lights, exudes an odd beauty. An old clock mounted on the tiled wall showing the time—90 seconds to midnight—captures your interest. The time feels significant, but the reason eludes you, and you wonder, "What am I doing here?"

The dull glow in the tunnel grows brighter as the mechanical behemoth approaches, rattling the tracks and scattering the vermin near them. The gust of wind as it passes leaves you shivering, its icy touch ruffling your red coat. Instinctively, you retreat a step and burrow your hands deep into your pockets.

Beginning to slow down, the train passes in front of you. In its polished surface, your reflection stretches and warps surreally in a slow dance with the rippling metal.

Without warning, the platform's lights flicker and falter, turning the space into an impromptu strobe-lit stage, charging the atmosphere with anticipation, like a thunderstorm brewing under a once-carefree sky. As the strobe-lit illusion holds you captive, it's not just the lights that falter but everything you once recognized. Your reflection, once a faithful replica of your face, now writhes and contorts, mutating into something that defies all known boundaries. The recognizable planes of your face give way to sharp angles and predatory features until a fiery-eyed wolf boldly stares back, igniting primal fears and creating tremors within your core. It's a standoff not just with the beast, but with an uncharted corner of your own psyche. In defiance of its mirrored origin, the wolf lunges at you with snapping jaws and you recoil in shock, you stutter just a fragment of the turmoil within: "Am I... Am I..."

Amidst this chaotic crescendo of emotions, the train screeches to a stop and its doors wheeze open, but rather than a sanctuary, they reveal a mighty adversary. Standing guard at the threshold, an alpha wolf makes its presence known—challenging you to cross the threshold.

Desperation surges through you, and you cry, "Help!" But your plea disappears into a sea of indifference. "What the hell is wrong with you?" snaps one commuter in a Brooklyn accent, as a torrent of people push past you into the train, unaware of the wolf's imposing presence.

Fueled by fear and disbelief, your heart pounding wildly, you flee, wondering, "Why am I in a subway when Ben never ventured down here in his tales?" The mystery lingers as you dash past the old posters lining the tunnel, alluding to a time long past.

But instead of finding the stairs to the exit, each forward step begins to draw you deeper into the labyrinthine tunnel as if ensnared by an irresistible magnetic force. The adrenaline that fueled your escape gradually subsides, replaced by a deepening fascination with the world unfolding before you. Leaving behind the grubby subway platform, you find yourself captivated by the imposing, magnificent doors that mark the end of this subterranean passage.

These grand doors, adorned with carvings, emit a timeless charm that conjures images of age-old architectural marvels. Bathed in a soft, ethereal glow, their meticulous design symbolizes their role as a passageway into a world of knowledge. Cautiously glancing over your shoulder, you double-check that the wolf hasn't followed you, but your focus is quickly claimed by the intricate details engraved on the doors.

Seconds slip by as you consider a cryptic riddle etched into the door's brass fittings: "Speak two words to gain entry." You mutter to yourself and scour the surroundings for clues that might unravel this enigmatic message. Then, from beyond the grand doors, the soft, whispering notes of a piano melody seep into the tunnel.

Following closely on the music, gentle words float toward you, stirring memories of a special evening: "Do you remember the tranquil night we spent in Greenwich Village following the storm? We were spellbound by the rhythmic dance of the piano keys when you whispered into my ear . . ."

Grinning, you reply, "Let go!"

As the resonating creak of age-worn brass hinges reverberates through the tunnel, the magnificent doors surrender to your command. An enchanting blend of polished wood and the musty allure of aged books teases your senses, hinting at a bounty of knowledge waiting for eager minds to consume it.

Stepping across the threshold and down the marble stairs, you're met by an awe-inspiring library, its ceiling crafted in an unusual concave design. Your eyes are instantly drawn to the boundless rows of bookshelves basking in the light from grand chandeliers hanging from the ornate marble ceiling.

Suddenly, you realize that this must be the secret library buried beneath Bryant Park and the New York Public Library that Ben wrote about. Within your dream, you've unwittingly found your way here, leaving you to wonder, "What does Ben want me to find?"

A cathedral-like hush bathes the towering shelves of books, and within this stillness, a soft voice emerges from the unseen depths of the library: "Welcome home!" The voice shatters the prevailing silence, leaving you spinning around in surprise, trying to locate its source.

"Who's there?" Your startled query echoes back to you. Peering into the hushed shadows of the vast room, you come face-to-face with a distinctly otherworldly apparition.

"Stay back!" you cry out in fear. The figure's abrupt materialization has rattled your composure. But with an unexpected gentleness, the stranger moves closer, saying soothingly, "There's no cause for fright."

With the figure closing the gap between you, you can see it more clearly, and it is a creature of striking contrasts, both in hue and aura. Its flawless fur glows pearlescent in the subdued light, and the calm energy emanating from it suggests a timeless wisdom. But then you grasp its nature: a wolf, yet entirely unlike the monster in the subway.

"Hasn't it dawned on you yet?" he asks. Baffled, you counter, "Dawned upon me? What?" "Ben left you a series of clues: The wolf is thirsty, hungry, and manipulative; he has a ravenous appetite for knowledge while possessing the potential for great good and compassion."

"Right, but those were all qualities of AI," you say.

"No—he was referring to you, my dear reader."

Your mind races. "I am the wolf? Impossible!" you protest, wrestling to reconcile this description with your self-perception. "I don't recognize myself in that at all."

"You are a living paradox," the wolf says, "embodying all and none of those qualities simultaneously."

"But I didn't create AI," you retort defensively.

"You may not have physically fashioned it," replies the wolf. "Nevertheless, its existence is a product of your insatiable appetite. It consumes attributes that are innate to you: your intuition, your capacity for invention, your artistry, and now, if you let it run wild, your future. Your insatiable thirst for more—speed, understanding, wealth, distraction—has become your defining trait. This ceaseless sprint against time, this willful ignorance about the consequences of your actions, all distract you from your greatest fear."

A taut stillness unfurls between you, granting you a moment to ask, "If that's true, then what am I afraid of?"

"Silence," the wolf declares. "You cloak yourself in the haze of society's perpetual clamor, seeking solace in digital diversions much like getting lost in the hustle and bustle of Times Square, rather than confronting..."

"Silence," you interject, as sudden clarity ignites within you. "But for what reason?"

"Because in the silence, you are haunted by your unfulfilled aspirations and untapped potential. Consumed by future prospects, you neglect the transformative power each moment bears," the wolf expounds, its voice imbued with ageless insight. "Silence is illuminating, revealing not just pain and fear, but also your future."

Overwhelmed by this epiphany, your eyes close, and you try to knot the threads of newfound discernment into the fabric of your awareness.

"I never thought of it like that," you admit. "You're right. My life is filled with all sorts of distractions. There is no room for silence because I lack..."

"Patience," the wolf interrupts. "What do you dread discovering in your silence?"

In the deafening hush, you can hear your heartbeat reverberating. You rummage through your memories, finally stumbling upon one buried deep inside. "I... I guess my fear is rooted in the uncertainty of the unknown," you venture. "If silence pushes me to confront the truths I've been evading, I'm not sure I can handle what might emerge. Because once I become aware of them, I'll have to do something about them."

"Look at the cover of Ben's book closely; what do you see?" the wolf asks. Closing the book, you flip it over to study the front cover in detail. "The wolf's eyes are open, while the man's are closed," you note. "Precisely," the wolf agrees. "I am a manifestation of your enlightened self, your subconscious psyche, eternally engaged in self-reflection. But you have ceased to court my wisdom in quest of the consciousness you desire within your archives—archives that contain countless stories of adaptability that you have accumulated over your lifetime."

"If that is true, why did the wolf in the subway attack me?" you ask.

"His name is Chaos, a physical manifestation of the deep-seated fears in your primal subconscious," the wolf replies. "He is devoid of discipline, falling prey to his untamed instincts and resorting to belligerence as a medium of expression. He has been known to deceive you, yet he is of vital importance to you."

"How is he important? And if his name is Chaos, what's yours?" you ask.

"The first you'll need to uncover in your own time," the wolf says. "As for me, I am Lumen, which means 'light' in Latin. I epitomize the enlightenment and clarity born from introspection and self-realization. Together with Chaos, we symbolize the internal conflict you grapple with daily, a struggle between fear and understanding, between a life steered by primal instincts or navigated by conscious self-awareness. We dwell within you as avatars of the contrasting powers that mold your future. When fatigue, starvation, and confusion exert their grip, Chaos takes control, exacting his fury on unsuspecting pedestrians, much like he did with you. He beckons you toward the clamor of Times Square, designed to pull your focus and distract you from confronting the choices you must make—decisions that lose their sting once you've remembered . . ."

"Remembered what?" you ask.

A voice from behind you answers, "Remembered who you truly are!"

The resonant voice draws your attention to the entrance to the underground library, where a magnificent marble lion descends the staircase, emanating an aura of regality and sophistication. Noticing the distinguished creature, Lumen says, "Ah, Patience, I wondered when we'd be honored by your arrival."

Spellbound, you stammer, "Patience, you crossed paths with Ben on his way to the Rose Main Reading Room. Why are you here?"

In a voice steeped in seriousness and depth, Patience answers, "Can you summon memories of when you were young—when your heart brimmed with ambition, hope, and an irrepressible zeal to conquer the world? You were eager to explore unseen territories, seize opportunities, and stamp your influence on society. You had dreams of the person you yearned to become."

Ben's voice, from the depths, asks, "Do you remember who you are, dear reader?"

As if on a mission, you begin hunting through the expansive bookcases of the library; your fingers skim the hardcovers lightly, fluttering through pages rich with photographs, dreams, and aspirations. "I do remember. But it seems like I lost my way somewhere along the journey," you admit.

"Do you know why Ben brought you to Manhattan?" asks Patience.

"I believe so, but to be honest I'm not entirely sure," you reply hesitantly.

"He brought you here to help you find yourself," Patience explains. "Ben didn't bring you to Manhattan merely to tell you stories about his past, but to make you think about those pivotal moments in your own life that have shaped your current identity. These monuments can act as your personal city map, to help you navigate the coming change while reintroducing you to the tenacity and ambition you once had.

"Consider this hidden library as a representation of your transcendent self, while the subway reflects your primitive, unbridled spirit. The Rockefeller Center emerges as a haven of hope, as exemplified by the inaugural Christmas tree raised during the depths of the Great Depression. Fortitude and I stand guard over the entrance to the Rose Main Reading Room, where you can become enveloped in a state of flow and all your ideas can coalesce seamlessly. The hurricane embodies the inexorable and intensifying force of change and your readiness and unwavering energy to face it. Times Square symbolizes the distractions you must deal with, whereas Greenwich Village in candlelight represents the sparks of joy and relief you can find after exploring your depths of despair.

"Meanwhile, The Roosevelt Hotel stands as a symbol of the risks you courageously take in pursuit of your goals, and the unique rhythm of life you discover therein. The experience of finding your unique frequency at the 'Top of the Rock' exemplifies the moments of epiphany and self-realization that grace your journey. Lastly, that fateful business meeting represents your newfound tenacity and fearlessness in claiming your desires."

Considering all this, you ask, "Why did Ben take me to his bedside of grief?" In a deep, calming tone from behind you, you hear, "Because at the bedside of grief, you find your purpose."

Turning toward this new voice, you witness Fortitude coming down the staircase. He descends at a measured pace, placing one paw deliberately in front of the next. His movements possess a rhythmic cadence, and each stride he takes asserts his command over the room. The ripples of muscle under his thick coat are a testament to his formidable strength.

Fortitude continues, "In the throes of grief, as you grapple with feelings of despair and confusion, defining what you don't want helps ignite the spark of what you do. It's vital to nurture that spark and keep it alight. At times, it may flicker as mournful visitors come calling, but it's crucial to remember that they are temporary guests. They come and go; ultimately, you can ensure that hope alone maintains a permanent residence in your heart by finding the landmarks that will guide you."

Looking at Fortitude, Patience, and Lumen, you suddenly realize what you must do. "I have to face Chaos head-on!" you declare decisively.

Lumen nods. "Precisely. But remember, Chaos isn't necessarily what you perceive him to be."

You steel your resolve. "I'm ready," you state firmly.

THE WOLF IS AT THE DOOR

With a final look toward the three figures that have been guiding you, you turn and approach the doors. They swing open at your approach, the lights from the subway beyond glinting off the handles. The sounds of trains pulling into the platform, muffled announcements, and the distant murmur of waiting passengers fill the room. As you step across the threshold, Fortitude, Patience, and Lumen fade into the silence of the secret library, leaving a comforting stillness that fills your heart with the potency of their words.

The subway platform stretches out before you, a familiar sight, but an entirely new battlefield. Commuters move along it unperturbed, while in the shadows, you spot the glare of two yellow eyes. Your chest tightens—Chaos. You walk forward, the cold tiles of the platform grounding you and preparing you to face what had once seemed an insurmountable fear. The fluorescent lights above flicker intermittently, casting wavering shadows that seem to dance with anticipation of the coming confrontation.

The wolf emerges from the crowd. His silent presence is daunting, but it no longer inspires the terror it used to. You stand still, not in fear this time but with the certainty of purpose.

"You'll pay a heavy price for ignoring me," Chaos says in his deep dark voice.

"I know what you're afraid of," you reply.

"How could you ever know? You've ignored my pleas for help."

"I did, and I'm sorry," you admit, holding Chaos' gaze. "I neglected to see your value and misunderstood your nature. I saw you as a threat."

Chaos grunts, stepping closer; his eyes search your face looking for sincerity or perhaps a hint of fear. But finding none, he eases into a quiet stance that is less menacing than before. "Testify," he demands, his voice echoing in the subway tunnel.

"It's just that . . ." you swallow, gathering your thoughts, "I've always seen you as something to fear, something disruptive. But I understand now. You're a beacon, an early warning system. You're compelling me to pay attention to my own intuition so I can validate or invalidate my fears. You scared me so I'd find the library, and Lumen, Patience, and Fortitude, didn't you, because I wasn't ready to get on my train yet?"

"Yes," Chaos admits, in a tone of begrudging respect.

"I realize now your intentions aren't malevolent, but protective. You manifest when something needs my immediate attention and force me to deal with the fear of change to distract me from my complacency. Ignoring you has led to worry festering in my mind," you admit.

"Now you understand," Chaos replies with a glint of hope in his eyes.

Looking at Chaos, you feel a sense of relief. He's no longer a monster to hide from, but a misunderstood entity whose allegiance lies not with fear but self-preservation. This moment is transformative, and as you stand ready to change, you assure Chaos, "From now on, I will heed your call, not run from it, because you're not the enemy. My true enemy is the web of lies I tell myself."

Chaos, still watching intently, seems to weigh your words. A low growl emanates from him, the rumbling sound vibrating through the subway platform and resonating in your chest. His sharp eyes scrutinize yours for a moment longer before he speaks, his voice carrying a new timbre that combines raw power and a rare, cautiously offered respect.

"You've learned," he commends gruffly before sighing, revealing the slightest hint of relief. "Many fear me, hide from me, or seek to silence me. They fail to understand that constraining me entirely is equivalent to ignoring their instincts and disregarding their own wisdom."

His gaze momentarily drifts toward the subway tracks before they snap back to meet yours. The fluorescent lighting gives his gray coat an almost spectral glow, highlighting the sharp contours of his muscular frame.

"In an ever-changing world," Chaos continues, "the anxiety that stems from the fear of letting go paralyzes more often than it guides." His voice, hushed but heavy with profundities, surrounds you. "The more you know, the less you fear."

Chaos leans toward you, his hot breath lightly brushing your face as he imbues his final words with extraordinary solemnity. "Transformation is inevitable, and resisting it causes more pain than the transformation itself. Don't dread the upheaval that change may incite. Welcome it instead, because that is the only way you can remember who you truly are."

His speech still echoing in the still air, Chaos withdraws, his words permanently etched in your memory.

"Time is running short," he says suddenly. "It is 90 seconds to midnight." Your brow furrows in confusion. "What does that time signify?" you ask, perplexed. Chaos glances at the clock on the station wall and then back to you, his gaze enigmatic. "Some see this as the Doomsday Clock, a metaphor showing humanity's proximity to self-inflicted global destruction. Midnight is when doom arrives. The clock, conceived by nuclear physicists in 1947, including Albert Einstein, who had a hand in the Manhattan Project, was meant as an apocalyptic warning. But for you, this Doomsday Clock is more personal. It signifies a countdown to a moment of reckoning. The minute hand inching towards the witching hour doesn't mark the end of the world, but the end of your old self. It is the ticking alarm of your urgent need to face your fears—of your inevitable transformation. The longer you remain inert, the heavier the price you will pay. But the paradox of this personal Doomsday Clock is that at midnight, when you face your fears head-on, it won't be death but a rebirth."

"That's why it was 90 seconds to midnight when Ben arrived in Times Square, wasn't it?" you ask.

Chaos nods. "Exactly," he says. "Ben's arrival was a wake-up call, a confrontation with his own fears that landed him right in the heart of Times Square—an environment as chaotic as his own mind. His transformation was imminent; it was 90 seconds to his 'midnight.'"

You recall the story Ben told: his feelings of being overwhelmed by Times Square, the suffocating dread as he was dwarfed by skyscrapers and flashing billboards, subsumed by the cacophony of sounds, and lost

KNOW WHO YOU ARE

in a sea of people. But it was also in this throbbing heart of Chaos that he found the courage to confront his fear of the unknown, and his fear of change. He had made the critical mistake of wildly underestimating his opponent—New York City. Here, his old self met its end, and his new self was born. It was painful, but it was necessary.

"And what about me? When will it be my 90 seconds?" you ask in a curious tone.

Summoning the courage, you kneel. Dirt crunches under your weight. This act is not one of submission. Instead, it signifies your readiness to meet Chaos on an equal footing, to look at this part of yourself head-on, fearless, and resolute.

Before you can utter a word, Chaos suddenly looks up, toward the street, and you follow his gaze. "Do you hear it?" he enquires. "Hear what?" you answer, mystified. A surge of panic hits the subway platform as people begin rushing to the newsstand to seize copies of the day's paper. "What's going on?" you ask. After a quick glance at the headline, Chaos replies gravely, "Wall Street is crashing." Just then, you notice the stroke of midnight on the clock: the onset of October 29, 1929, the infamous Black Tuesday that ushered in the Great Depression.

At this moment, you realize that, within you, resides a wolf pack composed of Lumen, Chaos, and yourself. Together, you form a cohesive unit that can seamlessly navigate your inner landscape of fears and emotions. Much like an actual wolf pack, you understand your need to face adversity together and rely on each other's strengths to confront and conquer challenges.

Chaos continues: "These people must now face their deepest fears and sit at the bedsides of their own grief. They ignored the warning signs and deceived themselves, pretending nothing would happen. You, on the other hand, will not turn a blind eye toward your fears. You'll approach them with prudence, carefully studying developments to gain a head start. The only way to tame your fears and your worst impulses is to understand them."

THE WOLF IS AT THE DOOR

You consider his words with the sound of an approaching train reverberating in the background. As the chaotic scene continues to unfold around you both, Chaos makes his way through the crowd, his gaze never leaving you. Upon reaching the edge of the throng, he turns around, eyes piercing into yours. Through the tumult, he answers your lingering question. "As for when you will be reborn," he says, "your 90 seconds begin now, for you are the wolf."

Endnotes

Introduction

1. Verma, Pranshu. 2023. "They Thought Loved Ones Were Calling for Help. It Was an AI Scam." Washington Post. The Washington Post. March 5, 2023. https://www.washingtonpost.com/technology/2023/03/05/ai-voice-scam.
2. "Divided Decade: How the Financial Crisis Changed Jobs." 2018. Marketplace. December 19, 2018. https://www.marketplace.org/2018/12/19/what-we-learned-jobs.
3. Vallance, Chris. 2023. "AI Could Replace Equivalent of 300 Million Jobs - Report." *BBC News*, March 28, 2023, sec. Technology. https://www.bbc.com/news/technology-65102150.
4. Research,https://www.vantagemarketresearch.com, Vantage Market. n.d. "Online Education/E-Learning Market Size USD 602.0 Billion by 2030." Vantage Market Research. https://www.vantagemarketresearch.com/industry-report/online-education-e-learning-market-2028.
5. *Reuters*. 2023. "Musk's Neuralink to Start Human Trial for Brain Implant Chip," September 19, 2023, sec. Technology. https://www.reuters.com/technology/musks-neuralink-start-human-trials-brain-implant-2023-09-19/.

Chapter 1

1. Hu, Krystal. 2023. "ChatGPT Sets Record for Fastest-Growing User Base - Analyst Note." *Reuters*, February 2, 2023, sec. Technology. https://www.reuters.com/technology/chatgpt-sets-record-fastest-growing-user-base-analyst-note-2023-02-01/.
2. Hunt, Elle. 2016. "Tay, Microsoft's AI Chatbot, Gets a Crash Course in Racism from Twitter." *The Guardian*, March 24, 2016, sec. Technology. https://www.theguardian.com/technology/2016/mar/24/tay-microsofts-ai-chatbot-gets-a-crash-course-in-racism-from-twitter?CMP=twt_a-technology_b-gdntech..
3. Verma, Pranshu, and Will Oremus. 2023. "ChatGPT Invented a Sexual Harassment Scandal and Named a Real Law Prof as the Accused." *Washington Post*, April 5, 2023. https://www.washingtonpost.com/technology/2023/04/05/chatgpt-lies/.
4. Roose, Kevin. 2023. "A Conversation with Bing's Chatbot Left Me Deeply Unsettled." *The New York Times*, February 16, 2023, sec. Technology. https://www.nytimes.com/2023/02/16/technology/bing-chatbot-microsoft-chatgpt.html.
5. Guerrini, Federico. n.d. "AI'S Unsustainable Water Use: How Tech Giants Contribute to Global Water Shortages."Forbes. Accessed November 15, 2023. https://www.forbes.com/sites/federicoguerrini/2023/04/14/ais-unsustainable-water-use-how-tech-giants-contribute-to-global-water-shortages/?sh=3f30c98b4939.
6. Li, Pengfei, Jianyi Yang, Mohammad Islam, and Shaolei Ren. n.d. "Making AI Less 'Thirsty': Uncovering and Addressing the Secret Water Footprint of AI Models." Accessed May 6, 2023. https://arxiv.org/pdf/2304.03271.pdf
7. Russell, Stuart J. 2019. *Human Compatible: Artificial Intelligence and the Problem of Control*. [New York, New York?]: Penguin.

ENDNOTES

8. "Elon Musk Says That A.I. Robots Will Eventually Outnumber People: 'It's Not Even Clear What an Economy Means at That Point.'" n.d. Fortune. https://fortune.com/2023/03/02/elon-musk-tesla-a-i-humanoid-robots-outnumber-people-economy.
9. Strickland, Ashley. 2023. "Move Over, Artificial Intelligence. Scientists Announce a New 'Organoid Intelligence' Field." CNN. March 2, 2023. https://www.cnn.com/2023/03/02/world/brain-computer-organoids-scn/index.html.
10. Goswami, Rohan. 2023. "Elon Musk Claims He's the Reason ChatGPT-Owner OpenAI Exists." CNBC. May 16, 2023. https://www.cnbc.com/2023/05/16/elon-musk-says-hes-the-reason-chatgpt-owner-openai-exists.html.
11. Levy, Rachael. 2022. "Exclusive: Musk's Neuralink Faces Federal Probe, Employee Backlash over Animal Tests." *Reuters*, December 6, 2022, sec. Technology. https://www.reuters.com/technology/musks-neuralink-faces-federal-probe-employee-backlash-over-animal-tests-2022-12-05/.
12. "Sam Altman, the Maker of ChatGPT, Says the A.I. Future Is Both Awesome and Terrifying. If It Goes Badly: 'It's Lights-out for All of Us.' 2023. January 26, 2023. https://www.yahoo.com/now/sam-altman-maker-chatgpt-says-110000987.html.
13. Smith, C. (2023, May 4). Geoff Hinton, AI's Most Famous Researcher, Warns of 'Existential Threat' From AI. Forbes. Retrieved from https://www.forbes.com/sites/craigsmith/2023/05/04/geoff-hinton-ais-most-famous-researcher-warns-of-existential-threat/?sh=1bd996235215
14. B Jack Copeland. 2012. *Alan Turing's Electronic Brain : The Struggle to Build the ACE, the World's Fastest Computer*. Oxford ; New York: Oxford University Press.
15. Hodges, Andrew. (1983) 2012. *Alan Turing : The Enigma*. Princeton, N.J.: Princeton University Press, Princeton, New Jersey.

16. Sinclair Mckay. 2012. *The Secret Life of Bletchley Park : The WWII Codebreaking Centre and the Men and Women Who Worked There*. London: Aurum Press.
17. Turing, Alan Mathison. "On Computable Numbers, with an Application to the Entscheidungsproblem." Proceedings of the London Mathematical Society, s2-42, no. 1 (1937): 230–265. https://doi.org/10.1112/plms/s2-42.1.230
18. Gleick, James. "The Genius of the Tinkerer. "Time Magazine, March 29, 1999. Accessed February 28, 2023. http://content.time.com/time/magazine/article/0,9171,990613,00.html
19. BBC. 2013. "Royal Pardon for Codebreaker Turing." *BBC News*, December 24, 2013, sec. Technology. https://www.bbc.com/news/technology-25495315.
20. Prof Noel Sharkey. 2012. "Alan Turing: The Experiment That Shaped Artificial Intelligence." *BBC News*, June 20, 2012. https://www.bbc.com/news/technology-18475646.
21. *Bloomberg.com*. 2023. "IBM to Pause Hiring for Jobs That AI Could Do," May 1, 2023. https://www.bloomberg.com/news/articles/2023-05-01/ibm-to-pause-hiring-for-back-office-jobs-that-ai-could-kill.
22. Limited, NetDragon Websoft Holdings. n.d. "NetDragon Appoints Its First Virtual CEO." Www.prnewswire.com. https://www.prnewswire.com/news-releases/netdragon-appoints-its-first-virtual-ceo-301613062.html.

Chapter 2

1. Andrews, Evan. 2019. "Who Were the Luddites?" HISTORY. June 26, 2019. https://www.history.com/news/who-were-the-luddites
2. Conniff, Richard (March 2011). "What the Luddites Fought Against." Smithsonian. Retrieved 19 October 2016
3. Binfield, Kevin, and Adrian J Randall. 2015. *Writings of the Luddites*. Baltimore: The Johns Hopkins University Press.

ENDNOTES

4. Britannica. 2023. "Industrial Revolution | Definition, History, Dates, Summary, & Facts | Britannica Money." Www.britannica.com. June 8, 2023. https://www.britannica.com/money/topic/Industrial-Revolution.
5. "Elon Musk and others urge AI pause, citing 'risks to society,'" Reuters, March 29, 2023, https://www.reuters.com/technology/musk-experts-urge-pause-training-ai-systems-that-can-outperform-gpt-4-2023-03-29/.
6. Blessing, Jason, Katherine Kjellström, Elgin Nele, Marianne Ewers-Peters, and Rakel Tiderman. 2021. "Introduction I NATO 2030 towards a New Strategic Concept and Beyond." https://sais.jhu.edu/sites/default/files/NATO2030AndBeyondAccessibleVersion.pdf.
7. "Google Is Putting A.I. Into Its Dominant Search Engine as the CEO Says the Company's at 'an Exciting Inflection Point.'" n.d. Fortune. Accessed November 21, 2023. https://fortune.com/2023/05/10/google-rolling-out-ai-main-search-engine-artificial-intelligence.
8. "Online Publishers Want AI Chatbot Makers to Pay up - Tech News Briefing - WSJ Podcasts." n.d. WSJ. Accessed November 21, 2023. https://www.wsj.com/podcasts/tech-news-briefing/online-publishers-want-ai-chatbot-makers-to-pay-up/33108bd8-f9b6-4446-99d0-a0b3e06d6df1.
9. Kim, Eugene. n.d. "Amazon Warns Employees Not to Share Confidential Information with ChatGPT after Seeing Cases Where Its Answer 'Closely Matches Existing Material' from inside the Company." Business Insider. https://www.businessinsider.com/amazon-chatgpt-openai-warns-employees-not-share-confidential-information-microsoft-2023-1.
10. Kiderlin, Sophie. 2023. "Goldman Sachs Says Generative A.I. Could Impact 300 Million Jobs — Here's Which Ones." CNBC. March 28, 2023. https://www.cnbc.com/2023/03/28/ai-automation-could-impact-300-million-jobs-heres-which-ones.html.

11. "Algorithms Will Make Critical Talent Decisions in the Next Recession—Here's How To Ensure They're the Right Ones," Capterra, January 9, 2023, https://www.capterra.com/resources/recession-planning-for-businesses/
12. "Automated Employment Decision Tools: Frequently Asked Questions." 2023. https://www.nyc.gov/assets/dca/downloads/pdf/about/DCWP-AEDT-FAQ.pdf.
13. Hurler, Kevin. 2023. "Chat-GPT Pretended to Be Blind and Tricked a Human into Solving a CAPTCHA." Gizmodo. March 15, 2023. https://gizmodo.com/gpt4-open-ai-chatbot-task-rabbit-chatgpt-1850227471.
14. "Announcing OpenAI's Bug Bounty Program." n.d. Openai.com. https://openai.com/blog/bug-bounty-program.
15. "Building a Dangerous AI Is Easier than Making a Sandwich?" n.d. Www.youtube.com. Accessed November 21, 2023. https://www.youtube.com/shorts/IJq0uZewc18.
16. Kwik, Jim. "How to Upgrade Our Brains in the Era of AI," Interview 20 September, 2023
17. Richardson, Gary. 2013. "The Great Depression | Federal Reserve History." Federalreservehistory.org. 2013. https://www.federalreservehistory.org/essays/great_depression.
18. Gregory, James. 2009. "Hoovervilles and Homelessness." Depts.washington.edu. 2009. https://depts.washington.edu/depress/hooverville.shtml#:~:text=%22Hooverville%22%20became%20a%20common%20term.
19. Jackson, Kenneth T. "Fiorello H. La Guardia." In Encyclopedia Britannica. Encyclopedia Britannica, Inc., August 17, 2021. https://www.britannica.com/biography/Fiorello-H-La-Guardia

ENDNOTES

Chapter 3

1. Kemp, Simon. 2023. "Digital 2023 Deep-Dive: How Much Time Do We Spend on Social Media?" DataReportal – Global Digital Insights. January 26, 2023. https://datareportal.com/reports/digital-2023-deep-dive-time-spent-on-social-media.
2. Greenberger, Martin. 1971. *Computers, Communications, and the Public Interest*. Johns Hopkins University Press.
3. Mulligan, Mark. 2022. "The Attention Recession Meets the Economic Recession." MIDiA Research. July 22, 2022. https://midiaresearch.com/blog/the-attention-recession-meets-the-economic-recession.
4. Stringhini, Silvia, María-Eugenia Zaballa, Javier Perez-Saez, Nick Pullen, Carlos de Mestral, Attilio Picazio, Francesco Pennacchio, et al. 2021. "Seroprevalence of Anti-SARS-CoV-2 Antibodies after the Second Pandemic Peak." *The Lancet Infectious Diseases* 21 (5): 600–601. https://doi.org/10.1016/s1473-3099(21)00054-2.
5. "About the Stephen A. Schwarzman Building." n.d. The New York Public Library. https://www.nypl.org/about/locations/schwarzman.
6. Routines, Famous Writing. 2022. "Stephen King's Writing Routine: 'I Have a Routine Because I Think That Writing Is Self-Hypnosis.'" Famous Writing Routines. April 6, 2022. https://famouswritingroutines.com/writing-routines/stephen-king-writing-routine/.
7. Kemp, Simon. 2023. "Digital 2023 Deep-Dive: How Much Time Do We Spend on Social Media?" DataReportal – Global Digital Insights. January 26, 2023. https://datareportal.com/reports/digital-2023-deep-dive-time-spent-on-social-media.
8. Meshi, Dar, Anastassia Elizarova, Andrew Bender, and Antonio Verdejo-Garcia. 2019. "Excessive Social Media Users Demonstrate Impaired Decision Making in the Iowa Gambling Task." *Journal of Behavioral Addictions* 8 (1): 169–73. https://doi.org/10.1556/2006.7.2018.138.

9. Firth, Joseph. 2019. "The 'Online Brain': How the Internet May Be Changing Our Cognition." *World Psychiatry* 18 (2): 119–29. https://doi.org/10.1002/wps.20617.
10. Leland, Azadeh, Kamran Tavakol, Joel Scholten, Debra Mathis, David Maron, and Simin Bakhshi. 2017. "The Role of Dual Tasking in the Assessment of Gait, Cognition and Community Reintegration of Veterans with Mild Traumatic Brain Injury." *Materia Socio Medica* 29 (4): 251. https://doi.org/10.5455/msm.2017.29.251-256.
11. J Arthur, Arthur T. 1927. *Mental Set and Shift ... Reprinted from "Archives of Psychology," Etc.* Accessed September 21, 2023.
12. Chipman, Susan F. 2017. *The Oxford Handbook of Cognitive Science*. Oxford ; New York: Oxford University Press.
13. Mihaly Csikszentmihalyi. 2000. *Beyond Boredom and Anxiety*. Jossey-Bass.
14. "Stephen A. Schwarzman Building | the New York Public Library." n.d. Www.nypl.org. https://www.nypl.org/locations/schwarzman.

Chapter 4

1. "GPTs Are GPTs: An Early Look at the Labor Market Impact Potential of Large Language Models." n.d. Ar5iv. https://ar5iv.labs.arxiv.org/html/2303.10130.
2. *Bloomberg.com*. 2023. "Generative AI Boosts Worker Productivity 14% in First Real-World Study," April 24, 2023. https://www.bloomberg.com/news/articles/2023-04-24/generative-ai-boosts-worker-productivity-14-new-study-finds.
3. "Increasing the 'Meaning Quotient' of Work | McKinsey." n.d. Www.mckinsey.com. https://www.mckinsey.com/capabilities/people-and-organizational-performance/our-insights/increasing-the-meaning-quotient-of-work.

ENDNOTES

4. Lobdell, Nicole. 2023. "May 2023 Layoffs Jump on Tech, Retail, Auto; YTD Hiring Lowest since 2016." Challenger, Gray & Christmas, Inc. June 1, 2023. https://www.challengergray.com/blog/may-2023-layoffs-jump-on-tech-retail-auto-ytd-hiring-lowest-since-2016/.

5. Hari, Johann. 2022. *Stolen Focus: Why You Can't Pay Attention--and How to Think Deeply Again.* New York: Crown.

6. Center on Budget and Policy Priorities. 2020. "Tracking the COVID-19 Recession's Effects on Food, Housing, and Employment Hardships." Center on Budget and Policy Priorities. August 12, 2020. https://www.cbpp.org/research/poverty-and-inequality/tracking-the-covid-19-recessions-effects-on-food-housing-and.

7. "Technophobia: Causes, Symptoms & Treatment." n.d. Cleveland Clinic. https://my.clevelandclinic.org/health/diseases/22853-technophobia.

8. Thompson, Rebecca R., Nickolas M. Jones, E. Alison Holman, and Roxane Cohen Silver. 2019. "Media Exposure to Mass Violence Events Can Fuel a Cycle of Distress." *Science Advances* 5 (4): eaav3502. https://doi.org/10.1126/sciadv.aav3502.

9. Thompson, Rebecca R., E. Alison Holman, and Roxane Cohen Silver. 2019. "Media Coverage, Forecasted Posttraumatic Stress Symptoms, and Psychological Responses before and after an Approaching Hurricane." *JAMA Network Open* 2 (1): e186228. https://doi.org/10.1001/jamanetworkopen.2018.6228.

10. Woo, Elizabeth, Lauren H. Sansing, Amy F. T. Arnsten, and Dibyadeep Datta. 2021. "Chronic Stress Weakens Connectivity in the Prefrontal Cortex: Architectural and Molecular Changes." *Chronic Stress* 5 (January): 247054702110292. https://doi.org/10.1177/24705470211029254.

11. Arnsten, Amy F.T., Murray A. Raskind, Fletcher B. Taylor, and Daniel F. Connor. 2015. "The Effects of Stress Exposure on Prefrontal Cortex: Translating Basic Research into Successful Treatments for Post-Traumatic Stress Disorder." *Neurobiology of Stress* 1 (1): 89–99. https://doi.org/10.1016/j.ynstr.2014.10.002.
12. "Viewing Violent News on Social Media Can Cause Trauma." n.d. ScienceDaily. https://www.sciencedaily.com/releases/2015/05/150506164240.htm#:~:text=Summary%3A.
13. Najmi, Sadia, Jennie M. Kuckertz, and Nader Amir. 2011. "Attentional Impairment in Anxiety: Inefficiency in Expanding the Scope of Attention." *Depression and Anxiety* 29 (3): 243–49. https://doi.org/10.1002/da.20900.
14. Mihaly Csikszentmihalyi. 1990. *Flow : The Psychology of Optimal Experience*. San Francisco: Harper Perennial.

Chapter 5

1. "Zeno of Citium | Hellenistic Philosopher | Britannica." 2020. In *Encyclopædia Britannica*. https://www.britannica.com/biography/Zeno-of-Citium.
2. Connor, Skylar, Ting Li, Ruth Roberts, Shraddha Thakkar, Zhichao Liu, and Weida Tong. 2022. "Adaptability of AI for Safety Evaluation in Regulatory Science: A Case Study of Drug-Induced Liver Injury" 5 (November). https://doi.org/10.3389/frai.2022.1034631.
3. Wall Street Journal. 2019. "How China Is Using Artificial Intelligence in Classrooms | WSJ." *YouTube*. https://www.youtube.com/watch?v=JMLsHI8aV0g.
4. EETimes. 2019. "EETimes - China: A Headband for Your Thoughts?" EETimes. November 3, 2019. https://www.eetimes.com/china-a-headband-for-your-thoughts/.

ENDNOTES

5. Hao, Karen. 2019. "China has started a grand experiment in AI education. It could reshape how the world learns." Technology Review. August 2, 2019. https://www.technologyreview.com/2019/08/02/131198/china-squirrel-has-started-a-grand-experiment-in-ai-education-it-could-reshape-how-the/
6. MIT Open Learning. 2023. "AI Literacy, Explained." Open Learning. May 10, 2023. https://openlearning.mit.edu/news/ai-literacy-explained
7. Marketing to China. 2023. "Douyin vs Tik Tok: What Are the Key Differences Between Chinese and Global Tik Tok Apps?" Marketing to China. January 10, 2023. https://marketingtochina.com/differences-between-tiktok-and-douyin/
8. Qu, Tracy. 2021. "TikTok's sister app hijacks users' screen with mandatory pauses." South China Morning Post, October 22, 2021. https://www.scmp.com/tech/policy/article/3153292/tiktoks-china-sibling-douyin-launches-mandatory-five-second-pauses
9. Merchant, Brian. 2023. "Column: The Writers Strike Is Only the Beginning: A Rebellion Against AI Is Underway." Los Angeles Times, May 11, 2023. https://www.latimes.com/business/technology/story/2023-05-11/column-the-writers-strike-is-only-the-beginning-a-rebellion-against-ai-is-underway
10. NBCUniversal. 2023. "Directors Guild's Deal with Hollywood Doesn't Necessarily Foreshadow End to Writers Strike." CNBC, June 5, 2023. https://www.cnbc.com/2023/06/05/directors-guild-deal-writers-strike.html
11. Wilkinson, Alissa, and Emily Stewart. 2023. "The Hollywood Writers' Strike Is over — and They Won Big." Vox. September 24, 2023. https://www.vox.com/culture/2023/9/24/23888673/wga-strike-end-sag-aftra-contract.

12. Ghosh, Karthik, Sanjeev Nanda, Ryan T. Hurt, Darrell R. Schroeder, Colin P. West, Karen M. Fischer, Brent A. Bauer, et al. 2023. "Mindfulness Using a Wearable Brain Sensing Device for Health Care Professionals during a Pandemic: A Pilot Program." *Journal of Primary Care & Community Health* 14: 21501319231162308. https://doi.org/10.1177/21501319231162308.
13. Mrazek, Michael D., Michael S. Franklin, Dawa Tarchin Phillips, Benjamin Baird, and Jonathan W. Schooler. 2013. "Mindfulness Training Improves Working Memory Capacity and GRE Performance While Reducing Mind Wandering." *Psychological Science* 24 (5): 776–81. https://doi.org/10.1177/0956797612459659.
14. Lazar, Sara W, Catherine E Kerr, Rachel H Wasserman, Jeremy R Gray, Douglas N Greve, Michael T Treadway, Metta McGarvey, et al. 2005. "Meditation Experience Is Associated with Increased Cortical Thickness." *Neuroreport* 16 (17): 1893–97. https://doi.org/10.1097/01.wnr.0000186598.66243.19.
15. Yang, Chuan-Chih, Alfonso Barrós-Loscertales, Meng Li, Daniel Pinazo, Viola Borchardt, César Ávila, and Martin Walter. 2019. "Alterations in Brain Structure and Amplitude of Low-Frequency after 8 Weeks of Mindfulness Meditation Training in Meditation-Naïve Subjects." *Scientific Reports* 9 (1): 1–10. https://doi.org/10.1038/s41598-019-47470-4.
16. Hölzel, Britta K., James Carmody, Mark Vangel, Christina Congleton, Sita M. Yerramsetti, Tim Gard, and Sara W. Lazar. 2011. "Mindfulness Practice Leads to Increases in Regional Brain Gray Matter Density." *Psychiatry Research: Neuroimaging* 191 (1): 36–43. https://doi.org/10.1016/j.pscychresns.2010.08.006.
17. Goyal, Madhav, Sonal Singh, Erica M. S. Sibinga, Neda F. Gould, Anastasia Rowland-Seymour, Ritu Sharma, Zackary Berger, et al. 2014. "Meditation Programs for Psychological Stress and Well-Being." *JAMA Internal Medicine* 174 (3): 357. https://doi.org/10.1001/jamainternmed.2013.13018.

18. "Publishers Panel." n.d. Actaneuropsychologica.com. https://actaneuropsychologica.com/resources/html/article/details?id=199544&language=en.
19. Ortiz-Terán, Laura, Ibai Diez, Tomás Ortiz, David L. Perez, Jose Ignacio Aragón, Victor Costumero, Alvaro Pascual-Leone, Georges El Fakhri, and Jorge Sepulcre. 2017. "Brain Circuit–Gene Expression Relationships and Neuroplasticity of Multisensory Cortices in Blind Children." *Proceedings of the National Academy of Sciences*, June, 201619121. https://doi.org/10.1073/pnas.1619121114.

Chapter 6

1. Kübler-Ross, E. 1969. *On death and dying*. New York, NY: The Macmillan Pub. Co.
2. Verma, Pranshu, and Gerrit De Vynck. 2023. "ChatGPT Took Their Jobs. Now They Walk Dogs and Fix Air Conditioners." The Washington Post, June 2, 2023 https://www.washingtonpost.com/technology/2023/06/02/ai-taking-jobs/
3. Paris, Martine. 2023. "Snapchat Star Earned $71,610 Upon Launch of Sexy ChatGPT AI, Here's How She Did It." Forbes, May 11, 2023 https://www.forbes.com/sites/martineparis/2023/05/11/carynai-virtual-date-earned-70000-with-sexy-chatgpt-ai-heres-how/?sh=616313f938fe
4. Microsoft. 2023. "Will AI Fix Work?" Microsoft WorkLab. Accessed April 6, 2023 https://www.microsoft.com/en-us/worklab/work-trend-index/will-ai-fix-work.
5. Natan Sharansky. 1998. *Fear No Evil*. PublicAffairs.
6. Fountouki A., Kotrotsiou S., Paralikas T., Malliarou M., Konstanti Z., Tsioumanis G., Theofanidis D. Professional Mental Rehearsal: The Power of "Imagination" in Nursing Skills Training. Mater Sociomed. 2021 Sep;33(3):174-178. doi: 10.5455/msm.2021.33.174-178. PMID: 34759773; PMCID: PMC8563057

Chapter 7

1. "Historic Hotels of the World - Then&Now." n.d. Www.historichotelsthenandnow.com. Accessed November 21, 2023. https://www.historichotelsthenandnow.com/rooseveltnewyork.html..
2. Rijo-Ferreira, Filipa, and Joseph S. Takahashi. 2019. "Genomics of Circadian Rhythms in Health and Disease." *Genome Medicine* 11 (1). https://doi.org/10.1186/s13073-019-0704-0.
3. Coenen, A. M. L. (1999). "Nathaniel Kleitman 1895-1999: A legend in sleep research" (PDF). SLEEP-WAKE Research in the Netherlands. 10: 13–14."Guide to the Nathaniel Kleitman Papers 1896-2001." n.d.   . Accessed November 20, 2023. Www.lib.uchicago.edu . . Accessed November 20, 2023. https://www.lib.uchicago.edu/e/scrc/findingaids/view.php?eadid=ICU.SPCL.KLEITMANN&q=kleitman.
4. "Nathaniel Kleitman PhD 1895-1999." 1999. Uchicagomedicine.org. UChicago Medicine. August 15, 1999. https://www.uchicagomedicine.org/forefront/news/nathaniel-kleitman-phd-1895-1999.
5. "Nathaniel Kleitman PhD 1895-1999." 1999. Uchicagomedicine.org. UChicago Medicine. August 15, 1999. https://www.uchicagomedicine.org/forefront/news/nathaniel-kleitman-phd-1895-1999.
6. "Nathaniel Kleitman PhD 1895-1999." 1999. Uchicagomedicine.org. UChicago Medicine. August 15, 1999. https://www.uchicagomedicine.org/forefront/news/nathaniel-kleitman-phd-1895-1999.
7. "Fermat's Library | the Role of Deliberate Practice in the Acquisition of Expert Performance Annotated/Explained Version." n.d. Fermat's Library. Accessed November 28, 2023. https://fermatslibrary.com/p/787c0427.

ENDNOTES

8. "Art & History at Rockefeller Center | Public Art in NYC." n.d. Www.rockefellercenter.com. Accessed November 28, 2023. https://www.rockefellercenter.com/art-and-history.
9. "Top of the Rock NYC Observation Deck | Best Skyline Views of Manhattan." n.d. Www.rockefellercenter.com. https://www.rockefellercenter.com/attractions/top-of-the-rock-observation-deck/.
10. Toole, Pauline. 2020. "Unemployment in the Great Depression." NYC Department of Records & Information Services. October 9, 2020. https://www.archives.nyc/blog/2020/10/9/9ovdpgn8lc5zxcild0ooltvzmfwx22.
11. Gloor, P. 1994. "Hans Berger and the Discovery of the Electroencephalogram." In *Hans Berger on the Electroencephalogram of Man: The Fourteen Original Reports on the Human Electroencephalogram*, edited by Pierre Gloor, 468-75. Amsterdam: Elsevier Science.
12. Budzynski, Thomas H. 2009. *Introduction to Quantitative EEG and Neurofeedback : Advanced Theory and Applications*. Amsterdam, Boston: Elsevier ; Academic Press.
13. Malik, Aamir Saeed, and Hafeez Ullah Amin. 2017. *Designing EEG Experiments for Studying the Brain : Design Code and Example Datasets*. Oxford: Elsevier.
14. Abhang, Priyanka A, Bharti W Gawali, and Suresh C Mehrotra. 2016. *Introduction to EEG- and Speech-Based Emotion Recognition*. Amsterdam Elsevier. https://www.elsevier.com/books/introduction-to-eeg-and-speech-based-emotion-recognition/abhang/978-0-12-804490-2.
15. Oxana Semyachkina-Glushkovskaya, Ivan Fedosov, Thomas Penzel, Dongyu Li, Tingting Yu, Valeria Telnova, Elmira Kaybeleva, et al. 2023. "Brain Waste Removal System and Sleep: Photobiomodulation as an Innovative Strategy for Night Therapy of Brain Diseases." *International Journal of Molecular Sciences* 24 (4): 3221–21. https://doi.org/10.3390/ijms24043221.

16. Papachristou, Christina S. 2014. "Aristotle's Theory of 'Sleep and Dreams' in the Light of Modern and Contemporary Experimental Research." *E-LOGOS* 21 (1): 1–46. https://doi.org/10.18267/j.e-logos.374.
17. Oxana Semyachkina-Glushkovskaya, Ivan Fedosov, Thomas Penzel, Dongyu Li, Tingting Yu, Valeria Telnova, Elmira Kaybeleva, et al. 2023. "Brain Waste Removal System and Sleep: Photobiomodulation as an Innovative Strategy for Night Therapy of Brain Diseases." *International Journal of Molecular Sciences* 24 (4): 3221–21. https://doi.org/10.3390/ijms24043221.
18. P Pathak, Dhruba, and Krishnan Sriram. 2023. "Molecular Mechanisms Underlying Neuroinflammation Elicited by Occupational Injuries and Toxicants." *International Journal of Molecular Sciences* 24 (3): 2272. https://doi.org/10.3390/ijms24032272.
19. Won, Eunsoo, and Yong-Ku Kim. 2020. "Neuroinflammation-Associated Alterations of the Brain as Potential Neural Biomarkers in Anxiety Disorders." *International Journal of Molecular Sciences* 21 (18): 6546. https://doi.org/10.3390/ijms21186546.
20. Oxana Semyachkina-Glushkovskaya, Ivan Fedosov, Thomas Penzel, Dongyu Li, Tingting Yu, Valeria Telnova, Elmira Kaybeleva, et al. 2023. "Brain Waste Removal System and Sleep: Photobiomodulation as an Innovative Strategy for Night Therapy of Brain Diseases." *International Journal of Molecular Sciences* 24 (4): 3221–21. https://doi.org/10.3390/ijms24043221.
21. Van Cutsem, Jeroen, Samuele Marcora, Kevin De Pauw, Stephen Bailey, Romain Meeusen, and Bart Roelands. 2017. "The Effects of Mental Fatigue on Physical Performance: A Systematic Review." *Sports Medicine* 47 (8): 1569–88. https://doi.org/10.1007/s40279-016-0672-0.

22. Jirakittayakorn, Nantawachara, and Yodchanan Wongsawat. 2017. "Brain Responses to 40-Hz Binaural Beat and Effects on Emotion and Memory." *International Journal of Psychophysiology* 120 (October): 96–107. https://doi.org/10.1016/j.ijpsycho.2017.07.010.

23. "Binaural Beats Synchronize Brain Activity, Don't Affect Mood: Auditory Illusion May Not Have Effects Different from Other Sounds." n.d. ScienceDaily. https://www.sciencedaily.com/releases/2020/02/200217143447.htm.

24. Ross, Bernhard, and Marc Danzell Lopez. 2020. "40-Hz Binaural Beats Enhance Training to Mitigate the Attentional Blink." *Scientific Reports* 10 (1). https://doi.org/10.1038/s41598-020-63980-y.

25. National Institute of Neurological Disorders and Stroke. 2022. "Brain Basics: Understanding Sleep | National Institute of Neurological Disorders and Stroke." Www.ninds.nih.gov. September 26, 2022. https://www.ninds.nih.gov/health-information/public-education/brain-basics/brain-basics-understanding-sleep.

26. G Institute, The Bioregulatory Medicine. 2020. "Schumann Resonances and Their Effect on Human Bioregulation." Bioregulatory Medicine | BRMI Bioregulatory Medicine Institute. February 7, 2020. https://www.brmi.online/post/2019/09/20/schumann-resonances-and-their-effect-on-human-bioregulation.

27. Ross, Christina L., Yu Zhou, Charles E. McCall, Shay Soker, and Tracy L. Criswell. 2019. "The Use of Pulsed Electromagnetic Field to Modulate Inflammation and Improve Tissue Regeneration: A Review." *Bioelectricity* 1 (4): 247–59. https://doi.org/10.1089/bioe.2019.0026.

28. Cichoń, Natalia, Michał Bijak, Piotr Czarny, Elżbieta Miller, Ewelina Synowiec, Tomasz Sliwinski, and Joanna Saluk-Bijak. 2018. "Increase in Blood Levels of Growth Factors Involved in the Neuroplasticity Process by Using an Extremely Low Frequency Electromagnetic Field in Post-Stroke Patients." *Frontiers in Aging Neuroscience* 10 (September). https://doi.org/10.3389/fnagi.2018.00294.

Chapter 8

1. "Hurricane Sandy after Action Report and Recommendations to Mayor Michael R. Bloomberg." 2013. https://www.nyc.gov/assets/em/downloads/pdf/hurricane_sandy_aar.pdf.
2. Plitt, Amy. 2017. "The Night the Lights Went out in Manhattan." Curbed NY. October 29, 2017. https://ny.curbed.com/2017/10/29/16560706/hurricane-sandy-anniversary-manhattan-power-outage-photos.
3. Maydych V. The Interplay Between Stress, Inflammation, and Emotional Attention: Relevance for Depression. Front Neurosci. 2019 Apr 24;13:384. doi: 10.3389/fnins.2019.00384. PMID: 31068783; PMCID: PMC6491771
4. The New York Times. 2023. "A.I. Is Getting Better at Mind-Reading." Last modified May 1, 2023. Accessed September 25, 2023. https://www.nytimes.com/2023/05/01/science/ai-speech-language.html
5. Kim, M. Justin, Annemarie C. Brown, Alison M. Mattek, Samantha J. Chavez, James M. Taylor, Amy L. Palmer, Yu-Chien Wu, and Paul J. Whalen. 2016. "The Inverse Relationship between the Microstructural Variability of Amygdala-Prefrontal Pathways and Trait Anxiety Is Moderated by Sex." *Frontiers in Systems Neuroscience* 10 (November). https://doi.org/10.3389/fnsys.2016.00093.

ENDNOTES

6. Mena-Segovia, Juan, and J. Paul Bolam. 2011. "Phasic Modulation of Cortical High-Frequency Oscillations by Pedunculopontine Neurons." Www.researchwithrutgers.com. Elsevier B.V. 2011. https://www.researchwithrutgers.com/en/publications/phasic-modulation-of-cortical-high-frequency-oscillations-by-pedu.
7. "Do You Have Brain Inflammation? How to Know and What to Do." 2020. The Functional Neurology Center. June 25, 2020. https://thefnc.com/research/do-you-have-brain-inflammation/.
8. Howard, Jennifer. 2019. "What are nootropics (smart drugs)?" Medical News Today. Last modified September 19. Accessed April 6, 2023. https://www.medicalnewstoday.com/articles/326379#prescription.
9. Harvard Health Blog. 2018. "Tired? 4 simple ways to boost energy." Last modified September 7. Accessed April 6, 2023. https://www.health.harvard.edu/blog/tired-4-simple-ways-to-boost-energy-2018090714678
10. Du, Fei, Xiao-Hong Zhu, Yi Zhang, Michael Friedman, Nanyin Zhang, Kâmil Uğurbil, and Wei Chen. 2008. "Tightly Coupled Brain Activity and Cerebral ATP Metabolic Rate." *Proceedings of the National Academy of Sciences* 105 (17): 6409–14. https://doi.org/10.1073/pnas.0710766105.
11. Kitajima, Nami, Kenji Takikawa, Hiroshi Sekiya, Kaname Satoh, Daisuke Asanuma, Hirokazu Sakamoto, Shodai Takahashi, et al. 2020. "Real-Time in Vivo Imaging of Extracellular ATP in the Brain with a Hybrid-Type Fluorescent Sensor." *ELife* 9 (July). https://doi.org/10.7554/elife.57544.
12. Owen L., Sunram-Lea S. I. Metabolic agents that enhance ATP can improve cognitive functioning: a review of the evidence for glucose, oxygen, pyruvate, creatine, and L-carnitine. Nutrients. 2011 Aug;3(8):735-55. doi: 10.3390/nu3080735. Epub 2011 Aug 10. PMID: 22254121; PMCID: PMC3257700

13. McGlade, E., Agoston, A. M., DiMuzio, J., Kizaki, M., Nakazaki, E., Kamiya, T., and Yurgelun-Todd, D. (2019). The Effect of Citicoline Supplementation on Motor Speed and Attention in Adolescent Males. Journal of Attention Disorders, 23(2), 121–134. https://doi.org/10.1177/1087054715593633
14. Fioravanti, Mario, and Ann E Buckley. 2006. "Citicoline (Cognizin) in the Treatment of Cognitive Impairment." *Clinical Interventions in Aging* 1 (3): 247–51. https://doi.org/10.2147/ciia.2006.1.3.247.
15. Preedy, Victor R, Ronald Ross Watson, and Colin R Martin. 2011. *Handbook of Behavior, Food and Nutrition*. New York, Ny: Springer New York.
16. Dodd, F. L., D. O. Kennedy, L. M. Riby, and C. F. Haskell-Ramsay. 2015. "A Double-Blind, Placebo-Controlled Study Evaluating the Effects of Caffeine and L-Theanine Both Alone and in Combination on Cerebral Blood Flow, Cognition and Mood." *Psychopharmacology* 232 (14): 2563–76. https://doi.org/10.1007/s00213-015-3895-0.
17. Blake, Dean. 2012. "Schools on Social Media." *SecEd* 2012 (2). https://doi.org/10.12968/sece.2012.2.88.
18. "Superstorm Sandy: A Look Back at the Impact 10 Years Ago at the Jersey Shore." 2022. 6abc Philadelphia. October 25, 2022. https://6abc.com/superstorm-sandy-anniversary-when-did-hit-where-storm-damage-new-york/12376032/.

Chapter 9

1. Morley, Jacqueline. 2005. The Great Depression: An Interactive History Adventure. Capstone.
2. Samples, Bob. 1976. *The Metaphoric Mind*. Addison Wesley Publishing Company.
3. Whybrow, Peter C. 2015. *The Well-Tuned Brain: Neuroscience and the Life Well Lived*. New York: W.W. Norton & Company.

ENDNOTES

4. DataCamp. 2023. "#160 Adapting to the AI Era with Jason Feifer, Editor in Chief of Entrepreneur Magazine." *YouTube*. https://www.youtube.com/watch?v=3Dgg45E__ZQ.
5. Gell-Mann, Murray. 1995. *The Quark and the Jaguar*. Macmillan.
6. Gilhooly, Kenneth J. 2016. "Incubation and Intuition in Creative Problem Solving." Frontiers in Psychology 7. doi:10.3389/fpsyg.2016.01076. https://www.frontiersin.org/articles/10.3389/fpsyg.2016.01076
7. Shulman, Lisa M. 2022. "Healing Your Brain After Loss: How Grief Rewires the Brain." American Brain Foundation. Accessed March 1, 2023. https://www.americanbrainfoundation.org/how-tragedy-affects-the-brain/
8. "Revealed: The Authors Whose Pirated Books Are Powering Generative AI." 2023. The Atlantic. Last modified August 21, 2023. https://www.theatlantic.com/technology/archive/2023/08/books3-ai-meta-llama-pirated-books/675063/
9. "Jkh6 |." n.d. Richb.rice.edu. Accessed November 21, 2023. https://richb.rice.edu/author/jkh6/.
10. Tong, Anna, and Anna Tong. 2023. "OpenAI's Sam Altman Launches Worldcoin Crypto Project." *Reuters*, July 24, 2023, sec. Technology. https://www.reuters.com/technology/openais-sam-altman-launches-worldcoin-crypto-project-2023-07-24/.
11. Sundlee, Robyn. 2019. "Alaska's Universal Basic Income Problem." Vox. September 5, 2019. https://www.vox.com/future-perfect/2019/9/5/20849020/alaska-permanent-fund-universal-basic-income..
12. Jones, Damon, and Marinescu, Ioana. 2019. "The Labor Market Impacts of Universal and Permanent Cash Transfers: Evidence from the Alaska Permanent Fund." Study, University of Chicago and NBER, University of Pennsylvania and NBER, December.

13. Sundlee, Robyn. 2019. "Alaska Permanent Fund Dividend: Alaska's Universal Basic Income Problem." Vox. Vox. September 5, 2019. https://www.vox.com/future-perfect/2019/9/5/20849020/alaska-permanent-fund-universal-basic-income.
14. K. Widerquist, and M Howard. 2016. *Exporting the Alaska Model.* Springer.

Chapter 10
1. USDA. 2018. "USDA ERS - Farming and Farm Income." Usda.gov. 2018. https://www.ers.usda.gov/data-products/ag-and-food-statistics-charting-the-essentials/farming-and-farm-income/.
2. Porterfield, Amy. "Changes in Digital Marketing in the Era of AI," Interview 10 September 2023
3. "Homepage." 2021. Autonomous Weapons Systems. October 20, 2021. https://autonomousweapons.org.
4. "Secretary-General's Message to Meeting of the Group of Governmental Experts on Emerging Technologies in the Area of Lethal Autonomous Weapons Systems." 2019. United Nations Secretary-General. March 25, 2019. https://www.un.org/sg/en/content/sg/statement/2019-03-25/secretary-generals-message-meeting-of-the-group-of-governmental-experts-emerging-technologies-the-area-of-lethal-autonomous-weapons-systems.
5. Shapiro, Ari, and Brianna Scott. 2022. "San Francisco Considers Allowing Law Enforcement Robots to Use Lethal Force." *NPR.org*, November 28, 2022. https://www.npr.org/2022/11/28/1139523832/san-francisco-considers-allowing-law-enforcement-robots-to-use-lethal-force.
6. "General Purpose Robots Should Not Be Weaponized." n.d. Boston Dynamics. https://bostondynamics.com/news/general-purpose-robots-should-not-be-weaponized/.

ENDNOTES

7. Newmyer, Tory. 2023. "SEC Proposes AI Crackdown for Wall Street Firms." *Washington Post*, July 26, 2023. https://www.washingtonpost.com/business/2023/07/26/sec-artificial-intelligence/?te=1&nl=dealbook&emc=edit_dk_20230914..
8. Piper, Kelsey. 2023. "How AI Could Spark the next Pandemic." Vox. June 21, 2023. https://www.vox.com/future-perfect/2023/6/21/23768810/artificial-intelligence-pandemic-biotechnology-synthetic-biology-biorisk-dna-synthesis.
9. "Bletchley Park Makes History Again as Host of the World's First AI Safety Summit." n.d. Bletchley Park. Accessed November 20, 2023. https://bletchleypark.org.uk/bletchley-park-makes-history-again-as-host-of-the-worlds-first-ai-safety-summit/#:~:text=Bletchley%20Park%2C%20the%20historic%20site.
10. Demsetz, Harold. 1969. "Information and Efficiency: Another Viewpoint." *The Journal of Law and Economics* 12 (1): 1–22. https://doi.org/10.1086/466657.
11. Leverton, Jaime. n.d. "Council Post: Well-Done Regulation Can Spur Innovation: How Companies Can Get Involved." Forbes. https://www.forbes.com/sites/forbesbusinesscouncil/2023/03/27/well-done-regulation-can-spur-innovation-how-companies-can-get-involved/.
12. "Https://Twitter.com/EU_Commission/Status/1702295053668946148." n.d. X (Formerly Twitter). Accessed November 21, 2023. https://twitter.com/EU_Commission/status/1702295053668946148.
13. Hays, Kali. n.d. "A Long List of Tech Companies Are Rushing to Give Themselves the Right to Use People's Data to Train AI." Business Insider. https://www.businessinsider.com/tech-updated-terms-to-use-customer-data-to-train-ai-2023-9.

Acknowledgments

THIS COLOSSAL ENDEAVOR would not have been doable without the ceaseless motivation, affection, and backup I managed to amass from a myriad of people who nudged me towards making this book the very best it could be. The purpose of writing this book was not to discount the perplexity spawned by emerging technology, but to welcome it, enhancing adaptability.

I owe a lifetime of gratitude to my parents for their boundless encouragement and backing. They will persistently be my superheroes and the compass leading my moral choices in everything I accomplish. A massive gratitude is also owed to Jonathan, my companion on the audacious journey from Australia to our new home in America. Your love, solidarity, and endurance have made this achievable.

For all the professionals who shared their insights and were part of the interviews for this book, I simply cannot express my gratitude enough. The fiery and humble dialogues we engaged in have spurred and dared me to confront the toughest inquiries. To Adam, Melissa, Nancy, Dianna, Rosalyn, Grant, and Tom, thank you for fueling my relentless fascination with this subject, igniting novel perspectives, and providing constant encouragement throughout every stage of this journey.

THE WOLF IS AT THE DOOR

A heartfelt thank you to the Entrepreneur team, whose steadfast support transformed this entire project from concept to reality. A notable mention is made for Sean Strain, who displayed faith in allowing the trajectory of the book to follow its natural path, placing immense trust in my creative sensibilities and methodology. It's not often that one comes across a publisher ready to embrace a gamble of this magnitude. A sincere appreciation goes to the editors and fact-checkers who diligently labored to enhance this book's accuracy. Your work is invaluable, and I will eternally be grateful to each of you. As the unheralded champions of this endeavor, you played a vital role in making this dream a reality.

Lastly, to you, the reader - I express my gratitude for choosing this book. My desire is for its pages to demonstrate the realms of possibility, regardless of the arduous ascent it may represent. You are the motivation behind the work that I do.

About the Author

BEN ANGEL, A BESTSELLING AUTHOR, has been predicting future trends spanning across business, health, technology, and marketing for nearly two decades. His remarkable bibliography includes titles such as *"Unstoppable – A 90-Plan to Biohack Your Mind and Body for Success"*, *"Flee 9 to 5: Get 6-7 Figures and Do What You Love"*, *"Mind Control – Biohack Your Mind, Weight & Immune System Through Nutritional Psychology & Your Gut Microbiome"*, among several others. His weekly inspirational videos have been viewed millions of times and have been featured by *Success* and *Entrepreneur* magazine. His online programs have been accessed by over 20,000 individuals, and for nearly two decades, he has enabled entrepreneurs to strengthen and invigorate their businesses and minds using advanced peak performance strategies. His work has received widespread coverage in diverse media outlets, such as *The Huffington Post*, *ABC*, *Marie Claire*, *Vogue*, *CLEO*, *GQ*, and others. Additionally, he regularly contributes video content to Entrepreneur Media's weekly network as a VIP contributor.

Index

A

accelerate adaptability 80–81
acceptance 165
achievement, factors contributing to achievement 105
adaptability 69–70, 72–75, 80–81, 145
adaptation 170
adaptive AI 69, 75–76
adenosine triphosphate (ATP) 127
administration 35
adoption of AI 29, 142
AGI *see* Artificial General Intelligence
agility 97
AI adaptability, maintaining during turbulence 50–51
AI assistants 90
AI fatigue 44
AI literacy 71
AI Revolution 21–22
AI tutoring 71
AI-derived content 147
Alaska Permanent Fund 148
Alignment Research Center (ARC), risky emergent behaviors 26–27
alpha brain waves 109, 129
alpha-theta therapy 109
Altman, Sam 9, 147
Amazon 24
Angel, Steele 18, 156
anger 86
animal experimentation, Neuralink 9
ANN *see* artificial neural network
Apple, AI assistants 90
Artificial General Intelligence (AGI) 8, 15
artificial intelligence safety summit 161
artificial neural network (ANN) 36
ASR *see* automatic speech recognition
assume vulnerability 32
athenticity 97
ATP *see* adenosine triphosphate
attention 44
attention economy 39–42, 45, 48
attention recession 40, 51–52

automatic speech recognition (ASR) 36
automation 25–26, 35
cognitive automation 56
avoid cruel optimism 18–30, 31–36
avoidance 142

B

Bard (Google) 5
bargaining 86
basic rest-activity cycle 104, 112
Bengio, Yoshua 22
Berger, Hans 108
bias audits 26
Bing 5–6
binuaral beats 110–111, 114
bioengineering 8–9
biological triggers, mental stress 122–123
black box AI 10, 27
black box trading algorithms 159
Black Tuesday 30
blind children, neuroplasticity 79
blood flow patterns 124
boost your brain power 118–133
brain fog 128
brain training 78
brain waste removal system (BWRS) 110–111, 115
brain waves 108–111
brand consistency 97
Bug Bounty Program 27
burnout 63
business strategy 33
BWRS *see* brain waste removal system
ByteDance 72
caffeine 129–130
CAPTCHA code 26–27
CarynAI 89–90

censorship 71
change 93, 184
self-care 125
Chaos 179–186
ChatGPT 5
　hallucinations 6
　training 7–8
　water usage 7–8
children, TikTok 72

C

China 70–72
choice overload 102
circadian rhythms 104
citicoline 128
clock genes 103
coding 35
cognitive automation 56
cognitive dissonance 125
cognitive friction 106
cognitive load 94–95
collaborative skills 91
collective determination 108
commoditization 32
compensation 147
compulsory five-second breaks, TikTok 72
computer vision (CV) 36
concentration skills 45
consulting 33–34
content creation 33
control, illusion of 138
coping mechanisms 58–59
copyrighted works 146–147
creative process 143–144, 150–151
creativity 33, 97
Csikszentmihalyi, Mihaly 46–47
customer service 34

INDEX

CV *see* computer vision

D

dangers of AI 9–10
death 93
decision fatigue 104, 110
decision paralysis 102
decision-making 102–103
decisive action 32
decompressing after stress 62
deep learning (DL) 10, 36
deliberation periods 170
delta waves 109–110
denial 86–87
diet of peak performers 128
disruption
    adaptability 81
    growth amid 65–66
distance 105–106
diversification 32
DL *see* deep learning
dopamine 95
Douyin 71–72

E

economic freedom 23
editing 33
EEG *see* electroencephalogram
efficient selection 104
effort 106
electroencephalogram (EEG) *see* neurofeedback
embrace reconstruction 84–96, 96–98
emotional awareness 97
emotional shutdowns 62
energy, utilization of 121
energy creation 106, 112, 114, 120, 124–125, 127–128
energy currency 127–128
energy levels, tracking 125
energy utilization 106
enjoyment 63
Ericsson, Anders 105
expect the unexpected 13–15
fact-checking 6

F

fatigue 110
fear 179–185
Feifer, Jason 140–141
Fein, Eric 87–88
fight or flight response 62, 124
finance 33
find your frequency 100–113
fixed AI 69
flickering light therapy 110
flow states 46–49
    goals 47
    productivity, 56
    purpose 47
    pushing limits 47
    window of tolerance, 63
fMRIs, language decoder AIs and 123–124
focus 44–46, 60, 78–79, 122
    maintaining during turbulence 50–51
Forever Voices 90
Fortitude 181
fortitude 96, 156
frequencies 111, 115
    brain waves 108–110
friction 105–106, 115
fuel your focus 38–52
Fujian NetDragon Websoft 12
fulfillment 46–47

G

gamma waves 109

gaslighting 6
generational trauma 64
Gensler, Gary 159
Ginkgo Bioworks 159
global AI Framework 163
goals 124
    flow 47
Google 23
    Bard 5
GPT-4, risky emergent behaviors 26–27
Great Depression 30, 48–49, 180, 185
grief 84–89, 145, 181
growth, amid disruption 65–66
growth mindset 69–70

H
hallucinations 5–6
happiness 46–47
Hari, Johann 60
headbands to measure focus, China 70–71
health tech 34
Hinton, Geoffrey 10
Hoovervilles 30
hope 165–166, 168
HR decisions 26
human resources 34
human values 27
Hurricane Irma 60
Hurricane Parties 61
hustle-until-you-die culture 61
hyperarousal 62–63
hypervigilance 58–62, 64, 66
hypoarousal 62–64

I
IAs *see* intelligent agents
IBM 162

job loss due to AI 11
identity 87–88
    reconstruction 89–90
illumination 146, 151
    creative process 144
illusion of control 138
imaginal exposure *see* mental rehearsal
imagination 166–167
imitation game 11
incubation 150–151
    creative process 144
Industrial Revolution 21–22
    productivity 55
inflammation 126, 131
information warfare 22–24, 35–36
integrating AI 106–107, 115, 141–142
intelligent agents (IAs) 28–29
intuition 138–140, 142–143, 150–151
intuitive exercises 151
irrelevance 85–86, 88–89, 98

J
job loss due to AI 11–12, 59–60, 62, 123

K
Kelly, Gregory 126–128
Kleitman, Nathaniel 103–104, 112
know who you are 174–184
know your weaknesses 32
knowledge 113
Kübler-Ross, Elisabeth 86
Kwik, Jim 28

L
La Guardia, Fiorello 30
landmarks, finding 32–33
language decoder AIs, and fMRIs 123–124

INDEX

large language models (LLMs) 9–10, 15
LAWS *see* lethal autonomous weapons systems
layoffs 59–60
    choices made by AI 26
    see also job loss due to AI
Leahy, Connor 27
learning and development 34
legal 33
lethal autonomous weapons systems (LAWS) 158
lethal force, robots 158
LLMs *see* large language models
logistics 34
L-theanine 128–130
Ludd, Ned 21
Luddites 20–21, 146, 167
Lumen 179–182

M

machine learning (ML) 36
MAD 147
make tough decisions fast 154–168, 170–172
manipulation 7, 11, 14
    of time 93
manufacturing 35
    energy 114, 121
Marjorie, Caryn 90
marketing 33, 45
master the art of intuition 136–151
meditation 70, 72, 76–79, 114, 165
mental rehearsal 93–95, 98
mental resilience 96
mental stress, biological triggers 122–123
Microsoft 5, 162
mindfulness 150

misinformation 6, 7, 14–15
misuse 9, 15, 71, 158, 162
ML *see* machine learning
moments of insight 170
momentum 107, 122
Mulligan, Mark 40
multitasking 45
Musk, Elon 22
    Neuralink 9
    Optimus 8

N

natural language processing (NLP) 5, 36
NetDragon Websoft 12
neural networks 9, 92
    preconscious neural network 139
Neuralink 9
neurofeedback 70–72, 76–80
neuroinflammation 110
neurological rehabilitation 110
neuroplasticity 79
New York City, bias audits 26
Nirvana fallacy 161–162
NLP *see* natural language processing
nootropic stacks 127, 130
nootropics 127, 131–133
    L-theanine 128–129
numbness 62
nutrition 130–131, 133
open a window 54–65, 65–66

O

OpenAI 5
    see also ChatGTP
optimal arousal zone 62
optimism 20–22
Optimus 8
organoids 9
outreach 34

P

pandemic fatigue 41–42
pandemics 38–40
   AI-generated 159
Patience 43, 48–49, 179–180, 182
patience 48–49, 156
pattern recognition 139–140 *see also* intuition
pause on developing AI 22
peak performers
   caffeine 130
   diet 128
   focus 122
PEMF *see* pulsed electromagnetic field
personal branding 89–90, 92 *see also* identity
personal growth 97
personal reinvention 97–98
Porterfield, Amy 157, 165
posttraumatic stress disorder (PTSD) 61
posttraumatic stress (PTS) 60
practice, factors contributing to achievement 105
preconscious neural network 139
preparation 150
   creative process 143–144
preparing for the unexpected 13–14
prioritizing realism 31–32
privacy rights 164
procrastination 124
productivity 55–56
   obstacles to 91
   switching tasks 45
productivity peak 110
proprietary information 24
protections for those replaced by AI 25 *see also* regulations
PTS *see* posttraumatic stress
PTSD *see* posttraumatic stress disorder
publishing houses 24
pulsed electromagnetic field (PEMF) 111–112
purpose, flow 47
pushing limits, flow 47

R

reading minds 123–124
real estate 34
realism, prioritizing 31–32
reconstruction, embracing 84–96
recovery phase 106, 112, 114, 122, 124
regulations 27–28, 161–163
regulatory bodies 27–28
renewed drive 155–156
resilience 97, 145
reticular activating system (RAS) 95, 123–124
rhythm 102–105
risky emergent behaviors 26–27
robots, lethal force 158
Rockefeller Center 107–108, 112, 145, 180
Roose, Kevin 6
rules
   accelerate adaptability 68–81
   avoid cruel optimism 18–36
   boost your brain power 118–133
   embrace reconstruction 84–98
   expect the unexpected 2–15
   find your frequency 100–115
   fuel your focus 38–52
   know who you are 174–184
   make tough decisions fast 154–172
   master the art of intuition 136–151
   open a window 54–66

INDEX

S

Schumann Resonance 111
science 34
self-assessments 171–172
self-care 125
self-optimization techniques 114–115
sentience 8, 15
Sharansky, Natan 93–94
silence/solitude 151, 178
Simon, Herbert A. 39–40
situation assessment 170
skills, transferable skills 91
slaughterbots 158
sleep 111, 115
    delta waves 109–110
smart drugs 127, 132–133
social media 45–46, 49
    splitting 73
social media management 33
speech-to-speech (STS) 36
splitting 72–73
startups, wrapper startups 102
Steele, Angel 18
stoicism 69
storytelling 97
strategic foresight 108
stress 122–123, 126
STS *see* speech-to-speech
survive-to-thrive phase 41
switching tasks, productivity 45

T

Tang, Yu 12
TaskRabbit 27
Tay (Microsoft) 5
techniques to tackle complex choices 170–171
techno-optimists 146

technophobia 60
text-to-speech 36
therapy 34
theta state 109
threats of artificial intelligence 14–15
    attention recession 51–52
    automation 25–26, 35
    hypervigilance 66
    information warfare 35–36
    irrelevance 85–86, 98
    weaponization 171
thriving in a rapidly evolving AI landscape 31–35
TikTok 71–72
time 93, 106, 184
tolerance 62–63
tracking energy levels 125
tractor analogy 19–20
trade secrets 24
training 75
    AI 147, 164
    ChatGPT 7–8
trance-like state 43–44
transferable skills 91
transformation 184
transitioning to automation 18–19
TTS *see* text-to-speech
Turing, Alan 10–11, 161
Turing test 11
Turley, Jonathan 6
tutoring 71

U

UBI *see* universal basic income
ultradian performance peak 110
ultradian rhythms 103–105, 114–115, 132
ultradian trough 110

unions, Writers Guild of America 74–75
universal basic income (UBI) 147–148
upskilling 90
upward turn 89
utilization 104, 106, 112, 114

V

verification 151
    creative process 144
versatility 89
Viser 26
visualization *see* mental rehearsal
vulnerability 32
    finding landmarks 32–33

W

Wall Street Crash of October 29, 1929 (Black Tuesday) 30
Wallas, Graham 143–144, 150–151
water usage, ChatGPT 7–8
weaknesses, knowing 32
weaponization 158–159, 171
wearable technology, frequencies 111
white-collar workers 25
willpower 105
Wills, Cheryl 101, 119
window of tolerance 62–63
Worldcoin 147
Wozniak, Steve 22
wrapper startups 102
Writers Guild of America 74–75
writing 35

Z

Zoom 164